PENGUIN BOOKS

Diddly Squat

Jeremy Clarkson began his writing career on the *Rotherham Adver-tiser*. Since then he has written for the *Sun*, the *Sunday Times*, the *Rochdale Observer*, the *Wolverhampton Express & Star*, all of the Associated Kent Newspapers and *Lincolnshire Life*. He was, for many years, the tallest person on television. He now lives on Diddly Squat Farm in Oxfordshire, where he is learning to become a farmer.

Diddly Squat

A Year on the Farm

JEREMY CLARKSON

PENGUIN BOOKS

PENGUIN BOOKS

UK | USA | Canada | Ireland | Australia
India | New Zealand | South Africa

Penguin Books is part of the Penguin Random House group of companies
whose addresses can be found at global.penguinrandomhouse.com

Penguin
Random House
UK

First published by Michael Joseph 2021
Published in Penguin Books 2022

016

Copyright © Jeremy Clarkson, 2021
Illustrations © Garry Walton at Meiklejohn, 2021

The moral right of the author has been asserted

Typeset by Jouve (UK), Milton Keynes
Printed and bound in Great Britain by Clays Ltd, Elcograf S.p.A.

The authorized representative in the EEA is Penguin Random House Ireland,
Morrison Chambers, 32 Nassau Street, Dublin D02 YH68

A CIP catalogue record for this book is available from the British Library

ISBN: 978–1–405–94653–7

This book is dedicated to Kaleb, Charlie, Ellen, Kevin, Gerald and, of course, Lisa.

Contents

CONTENTS

Introduction

For more than twenty years I have written about cars in the *Sunday Times*. And then along came the bat flu, which meant the car-makers were forced to shut down their demonstration fleets. Which, in turn, meant that I had nothing to review.

Of course, I could have put my feet up, and instead of writing a car column every week, I could have written nothing at all. But I like writing. It's what I do to relax. Some people enjoy playing golf or doing the downward dog or walking up hills. I like sitting at my laptop, dreaming up similes and metaphors. So, I asked my editor if it would be all right to write about farming instead. And he said yes, and now, what we've done is put all those columns in a book.

It's strange. When I write about cars, I sort of know what I'm talking about. I know what understeer is and what a carburettor does. But I didn't know anything about farming. Literally nothing at all. Here I was then, the *Sunday Times* farming expert, and I could not tell barley from wheat, let alone how you made them grow. Nor did I know what rape was used for.

I certainly didn't understand any of the buttons on my tractor, and when I was told I'd need somewhere to store my grain, I didn't know whether I'd need a bucket, or a bath. It turned out I'd need a barn, and a big one at that.

All I could do then, is take you, the reader, on my voyage of discovery. And what a voyage it's been.

Since I began, I've learned that it is completely impossible to attach anything to my tractor and that the weather will always do something you weren't expecting, and don't want. I also learned that whatever you plan on doing with your day on the farm you will invariably end up doing something else, that sheep are an expensive nuisance, that wasabi isn't commonly grown in the UK for a reason, and that the government is endlessly annoying.

Most of all, though, I have learned that it is a great life. Being on my tractor when the sun is going down, watching the hares and the deer and the songbirds, is something I never knew I'd find so utterly joyful. And then stopping, briefly, at lunchtime to eat a sandwich made from my wheat and my lamb, and washing it down with my apple juice: well, it sure as hell beats watching James May organize his tool kit.

This morning, I have to top my hay and then, tomorrow, I'll be starting my second harvest, which means I probably won't get much of a chance to drive the Porsche GT3 that's sitting in the farm yard, waiting to be reviewed. Still, not the end of the world, eh?

SPRING

May

Mowing my meadow with a Lamborghini tractor

Back in 2008, I bought a thousand-acre spread in Oxfordshire and employed a local man to do the farmering. But last year he decided to retire, so I thought I'd take over myself. Many people were surprised by this, as to be a farmer you need to be a vet, an untangler of red tape, an agronomist, a mechanic, an entrepreneur, a gambler, a weather forecaster, a salesman, a labourer and an accountant. And I am none of those things.

My bosses at Amazon were so surprised, they commissioned an eight-part show that would enable viewers to enjoy the 'hilarious consequences' of my attempts to manage the woods and the meadows and the fields full of wheat and barley and oilseed rape. I'd called the farm Diddly Squat because that's what it makes.

Still, I was confident I'd manage. Man has been farming for 12,000 years, so I figured it must be in our DNA by now. You put seeds in the ground, weather happens and food grows. Easy.

Unfortunately I could not have picked a worse year to begin. We had the wettest planting season on record. It started raining in October and did not stop for seven weeks. Then there was the uncertainty about Brexit. And then, just as the sun came out, everyone was told to go indoors and stay there, possibly for ever.

This has had a catastrophic effect on prices. When I first began delivering my 140 lambs a couple of weeks ago, they were worth £100 each. Now that's down to £30. Spring barley, meanwhile? It'll be hardly worth harvesting, thanks partly to a weather-driven glut and partly to the fact that barley makes beer. And all the pubs are shut.

Despite the problems, however, I'm sitting here on a lovely spring day and, apart from 10 acres of oilseed rape eaten by flea beetles, everything seems to be growing quite well. And only three lambs have died. And as there's so much to do, I'm not wandering around the house, glugging wine from the bottle and watching reruns of *Cash in the Attic*. I'm a key worker.

And better yet, I still have something to write about, here in the motoring section of your newspaper – my tractor.

I could have bought a Fendt. Everyone says they're the best. Or I could have bought a Fastrac, because I'm friends with the JCB family. But obviously I wanted a Lamborghini. So that's what I've got. An R8 270 DCR, to be precise.

Lamborghini was a tractor-maker long before it made cars, but the business was sold – along with the rights to the name – in 1973. Today they're made in Germany but they still look Lambo-mad. If an Aventador were to make love to a spaceship, this is what you'd end up with.

It's huge. Even the front tyres are taller than me. You have to climb up a four-rung ladder to reach the door handle and then you climb up some more to get into the cab, and then up again to get into the seat. It's so vast, in fact, that it wouldn't fit into my barn. I therefore had to build a new one. Every single farmer type who's seen it says the same thing. 'That,' they intone with a rural tug on the flat cap, 'is too big.' But in my mind tractors are like penises. They cannot be too big.

Yet the farmers are quite right. It is too big. Not only will it not fit into my barn, it won't fit through the gate on to my driveway. So I've had to build a new driveway. It is also too powerful. The straight-six turbocharged diesel produces only 270 horsepower, which, in car terms, is Golf GTI territory, but there are 775 torques. This means that when you attach a piece of equipment to its rear end, it is immediately ripped to shreds.

Not that I can attach anything to its rear end. It's all heavy engineering back there and I just know that if I tried, you'd be reading about yet another farmer walking for four miles across his fields with his severed arm in a bag. To put cultivators and rollers and drills on the back, I've therefore had to employ a man called Kaleb. Who also says my tractor's too big. He reckons his CLAAS is better. We argue about this a lot.

I concede the Lamborghini is a bit complicated. You start it and there's an almighty roar from the vertical smokestack, which is a full 7ins in diameter. And then you put it in gear. And then you put it in gear with the other gear lever, and then you let the clutch in, before you realize you haven't selected forward from the other gear lever. To change gear on the move, though, you use a fourth gear lever.

There are, I'm told, forty-eight gears forward and reverse. Happily, there are only two brake pedals and two throttles. But I did count 164 buttons before I opened the arm rest and found 24 more. None of them is labelled, which is a worry as all of them are designed to engage stuff that will tear off one of my arms.

Eventually, though, it all began to move and I discovered something unusual. The tractor has suspension and so does the seat, but they are designed to operate independently, so when the tractor is going up, the seat is always coming down. This means you alternate between severe spinal compression and a banged head. I clung so desperately to the steering wheel that after just three minutes it came off. Literally, off.

I've never been terrified at 25mph before, but in that tractor I really was. Since then, I've driven it very slowly . . . into six gates, a hedge, a telegraph pole, another tractor and a shipping container. I think I'm right in saying that I have not completed a single job without having at least one crash. Doing a three-point turn at the end of a cultivating run? I'm bad at that. I always go through the fence.

I'm also very bad at 'drilling'. This is the word we farmer types use for 'planting'. Mainly this is because, to do it properly, you must install the type of computer that Nasa uses for calculating re-entry angles. That's another aspect of farming I can't do: computer programming. Which is why some of my tramlines are 10ft apart and some are in Yorkshire.

However, despite all this, when I'm trundling along and the air-conditioning is on and there's a constant dribble of socialism coming from Radio 4, I confess I start to understand why Forrest Gump was happy, after all his adventures, to end up on a tractor mowing the school football field. I'm especially happy when the engine is under load, because the stupendous noise coming from that exhaust pipe drowns out Marcus Brigstocke.

And when I finish a field and I climb down the ladder and sit on a fence I've just broken to enjoy a bottle of beer and a chicken sandwich, I can look back at the work I've done and feel a tiny bit proud. It's not nursing or doctoring, I understand that, but growing bread and beer and vegetable oil is somehow a damn sight more rewarding than driving round corners while shouting.

As I am not able to write columns about cars until this virus issue is solved, I shall be bringing you more news from the farm each week.

Cost of tractor, second-hand from Germany: £40,000

Cost of barn to put it in: £28,000

Cost of driveway it can actually use: £23,000

Cost of man to fit things to it every morning: His business, not yours

Cost of repairing the damage I've done so far: £215,000,000

But it does run on red diesel.

The vegetable crisis

There was a brouhaha recently about a planeload of Romanians who had arrived here to pick vegetables. 'We don't want their diseases,' said people in tracksuits. 'And why can't the jobs be given to proper English people?'

Hmm. Farmers have been screaming for weeks about how their vegetables will die unless an army can be raised to pick them. They've been begging 'proper' English people to get off their flabby arses and help out, but apart from a few middle-class parents who've signed up Giles for a week on his hands and knees, the response has been pathetic. There were 90,000 jobs on offer; 6,000 people got as far as an interview. Hence the plane from Romania.

Ordinarily I would not be interested in this story, because my farm is on a hill in the Cotswolds. When I come here from London, the temperature gauge in my car drops like the altimeter in a crashing airliner. It's cold here. Bitter. And that's the wrong weather for veg.

I'm also informed that the soil's no good. 'It's brash,' say the locals who wear overalls and Viyella shirts for a living. Many also wear ties. I'm not sure why. A tie is just something else that can get caught up in farm machinery. But anyway, they say 'brash' is good only for cereal crops. And maybe sheep. Not vegetables.

Last year, to prove them wrong, I decided to plant a couple of acres of potatoes. Eventually, after filling in a stack of forms about 4ft high, the government gave me permission (in farming, you have to get permission from the government to get up in the morning) and four months later I had 40 tons of spuds in the shed. This was the wrong amount: not enough to make it worth a merchant's while to send a lorry, too much to sell at the side of the road. I managed to sell one ton; 38 tons rotted; and I've given the rest away to old people in the village.

Financially, then, my attempts to become the potato king of Chipping Norton ended in failure. But it did prove to the locals you can grow vegetables up here in the freezing troposphere, in soil that's nine parts stone and one part dust.

That's why, a couple of weeks ago, I decided to take half a field earmarked for spring barley and use it instead to grow broad beans, beetroots, leeks, cabbages and all the other things people use as an accompaniment to food.

This meant buying a planting machine. Most, these days, are designed for planting a whole county in a morning and Canada by nightfall. But I had only a 4-acre plot, so I ended up buying one from the Fifties. It's tiny. And brittle. If I attached it to the back of the gigantic torque mountain that is my Lamborghini tractor it would explode. I therefore needed a smaller tractor. So I cleverly bought my girlfriend, Lisa, a present. It's a dinky little 1961 Massey Ferguson.

But despite my ingenuity, there was a problem. You need someone to drive the tractor and two people to sit on little chairs in the miniature planting machine, feeding the vegetable sets – as the seedlings are called – into the machinery. And there is no way that's possible when everyone has to be 6ft apart.

I called my children, who, despite the lockdown, immediately decided they had work to do and essays to write. So I got my tractor driver, Kaleb, to sit on the Massey Ferguson and we decided he was very nearly 6ft in front of the planting machine. Therefore Lisa and I could sit in it, doing the work.

It's said that deep-sea diving off an oil rig is dangerous work and soldiering is worse. But the fact is the fatality rate among people in agriculture is almost twenty times higher than the average for all industries. And when you sit in a planting machine you can see why.

In front of you, mounted vertically, is a heavy motorcycle-style chain, and attached to it, every 4 or 5ins, are little V-shaped platforms on to which you place the vegetable plants. As the tractor goes along, the chain turns and you start to get an idea of what it might be like to be inside a gearbox. It is phenomenally easy to get your hand trapped. And because the tractor is so loud, its driver would not hear your screams.

The planter is fitted with a plastic cover. Initially, I thought it was to shield the occupants from the sun and rain. Now I'm fairly sure it's to make life easier for the coroner.

The most amazing thing, though, is that the machine doesn't work. It either buries the sets a foot down where there's no sunlight or it doesn't bury them at all. This means you have to go over the ground you've covered and do it all again by hand. Until eventually you realize it's easier to plant everything by hand in the first place.

So that's what we did. Planted by hand, for hour after back-breaking hour. And for what? So some spoilt little fat kid can push the fruits of our labours to the side of his plate and demand a Twix instead.

Ha. Chance'd be a fine thing. We are not experts in market gardening. We aren't even on the bottom rung of the market gardening ladder, but even we were able to deduce, the day after we'd planted the first acre, that something was wrong. Our new plants were kind of leaning over. 'Wilting', I believe, is the correct word.

It turned out they needed water. And how do you get water to a field that's half a mile from the nearest tap? Well, you need a digger, a pipe-laying machine, a dam across one of the streams and a pump, and after you've done all that, a couple of men to come along and do it all again. Only properly. At this rate, the only way I can achieve profitability is by charging £140 for each broad bean. And £400 for a cabbage.

And that doesn't factor in the amount of time I'm giving to the project. Which is all of it. Ten times a day I move my four sprinklers to new positions – and they are running constantly, demanding so much water from the stream that there's very little left to supply my house. Most days I feel like Jean de Florette.

I woke yesterday to the sound of rain and for the first time in my life I was glad. But now it's sunny and windy and the forecast says it will be 24°C by the end of the week – 24°C in effing spring. After the wettest autumn on record. How come no one has noticed this sort of thing is happening?

The weather, however, is not my biggest issue. That'll come in the summer, when the vegetables that haven't died will need picking. If I use Romanians, Nigel Farage and his Hackett army will go berserk, and if I use Lisa's daughter, who's keen, the *Daily Mail* will accuse me of employing child labour. So it'll be down to me.

It'll kill me for sure. I'll become a farming statistic. But I guess I'll be able to crawl through the Pearly Gates knowing that I have the gratitude of Joan Armatrading, Jeremy Corbyn, Lewis Hamilton, Paul McCartney, Captain Sensible, Miley Cyrus and all the other celebrities who've chosen to follow in the footsteps of Adolf Hitler and lead a meat-free life.

They think they are being kind. But they aren't. Because eating vegetables is bloody cruel to the people who have to grow the damn things.

Sheep are vindictive. Even in death

Last week one of my pregnant sheeps watched one of its mates give birth and decided that the new and very slimy lamb was hers. So, much to the distress of the actual mother, she started to lick it and offer up her nipples – is that the right word?

Whatever, women tell me that the birth process is something they tend to remember. So how could a sheep think it had given birth when it hadn't? There's an obvious answer. Sheeps are the stupidest animals on God's green earth. Except for one thing. They're not.

I bought mine last year at an auction in Thame, Oxfordshire. I had no idea what I was doing. Sheeps were brought into the ring, the auctioneer made machine-gun noises and I went home with sixty-eight North Country Mules. I've no idea what I paid. I couldn't understand a word anyone said.

I then bought two rams, which are basically woolly ball sacks, and in short order, all but three of my new flock were pregnant. The failures? I ate them, and they punished me for that by giving me heartburn.

And this is what I've learnt about sheeps in the nine months I've had them. They are vindictive. Even in death.

Sheeps know that human beings are squeamish. As a result, they never die of something simple, such as a heart attack or a stroke. No. A sheep's death has to be revolting. So they put their head in a bit of stock fencing and then saw it off. Or they decide to rot, from the back end forwards. Or they get a disease that causes warts to grow in their lambs' mouths. A sheep's death has to be worthy of a Bafta. Remember Alec Guinness at the end of *The Bridge on the River Kwai*? Well, it's that. With added haemorrhagic enteritis.

My sheeps clocked me immediately as a chap who's eaten too many biscuits, so when I had to move them out of one field into another, they'd do exactly as they were told. Then they'd wait for me to close the gate and walk home, before jumping over the wall, back into the first field. Did you know they can jump? Well, trust me on this: if a sheep wanted to annoy you, it could win the Grand National.

I bought a drone eventually and programmed the onboard speaker to make dog-barking noises. This worked well for a day, but then the sheeps just stood there, staring at it. So I had to move them by running about. And as I trudged home with a bit of lung hanging out of my mouth, they jumped over the wall again.

Today I have 142 extremely delicious-looking lambs boinging around in the fields. The walkers still won't put their wretched dogs on leads but at least they now look guilty when I glower at them. Although, actually, the biggest problem is not the dogs. It's the mothers.

Last week one of them decided that, to annoy me, it would abandon its lamb. I found the poor little thing in a hedge, shivering and hungry, and any attempt to reunite it with its mother ended with the lamb, and me, on our backs. The ewe was having none of it.

So I had to bring the lamb to the barn and make a bed for it near the wood-burning stove and sit up all night with bottles of warm milk. And then, in the morning, because it's a sheep and it wanted to upset me, it died.

The only good news about this is that there's no financial loss. Owing to the double whammy of Brexit and Covid-19, lambs today are worth about the same as a barrel of oil – minus £30.

Still, at least I now know how it must have felt to be a guard at Stalag Luft III. Because what those sheeps are doing when they're standing there in a perfectly nice field is thinking of ways to escape. If they were people, they'd be Gordon Jackson, Charles Bronson and Steve McQueen.

They constantly probe for any weakness in the fences. They keep tabs on my routines. And I'm bloody sure they are imperceptibly turning one of the cross-country fences into a rudimentary vaulting horse. And it's not because they want to get out. They're in the best field with the best grass. They just want to get on to the road so they can be hit by a bus, and burst.

Their latest game is very irritating. Somehow they've worked out how to open the doors on the hen houses. Even though I have opposable thumbs, I can barely do this; the latches are very stiff. But they can. And at night, they do. This means the hens can escape, and that means they are killed by nature's second most vindictive animal – the fox.

I cannot work out why the sheeps open the doors. It's not as if they're after the eggs, or the hens. Which means they must be doing it for sport. They actually enjoy watching the hens being eaten. And, as an added bonus, it pisses me off, which they enjoy even more.

It's the same story with their water bowser. They've worked out how to break the tap so all the water leaks into the soil. This means that either I have to mend it, or they die of thirst. So for them, it's a win-win.

Last night they gnawed through the wire providing power for the electric fence. So they could get out? Nope. So I'd have to stop what I was doing and fix it.

As I was doing that, I noticed something odd about one of the lambs. Its ears had come off. And as I stood there with my hands on my hips, asking myself how that was even possible, I got a pretty good idea of what life was like for my teachers having to deal with me and my troublesome friends. 'Why have you rubbed linseed oil into the school cormorant, Clarkson?'

That's what sheeps are, I've decided. Woolly teenage boys. And that's why they are so annoying.

SUMMER

June

Stow the chainsaw, I've called in Godzilla

It's been suggested that after months of house arrest, people are starting to appreciate the countryside. Many are saying they won't need cities or beaches in future because there's so much to enjoy in an English wood.

I'm not sure my girlfriend, Lisa, subscribes to this point of view. In a previous life she had homes in the Swiss Alps, Mallorca and London. She sailed the Atlantic on a whim and then lived on a beach in Trinidad for a while. Her butler had a butler. Remember that Learjet that crashed on to the A40 in west London a while back? She was on it.

But as I write, she's in the vegetable field, wearing tracksuit bottoms that are too short, furry Ugg boots and a vest that is soaking wet because she tried to move the sprinklers in a howling gale, and now she's on her hands and knees poking bits of horse manure into the mud, desperately hoping this will help the leeks and golden beetroots fight off the constant bombing raids from the beetles, birds and mildew.

Alan Titchmarsh told *The Times* recently that all plants

want to grow. 'It's just up to us not to get in the way,' he said. That, Alan, is bollocks.

Lisa loved the idea of growing vegetables on the basis that all the work was done by a little man from the village and all she had to do on a summer's evening was wander around with a watering can and a trug. But because she's now been exposed to the brutal reality of full-on vegetable farming, I suspect that ninety minutes after the restaurants reopen, she will be in Sloane Square, powering through the door of the Colbert café like a 6ft artillery shell.

I suspect I shall be with her because, at the weekend, I decided to have a go at woodland management. It turns out, however, that you can't 'have a go' at this, any more than you can 'have a go' at underwater oil-rig maintenance.

All men believe that they can operate a chainsaw. And, furthermore, all men want to operate a chainsaw. Because it's the manliest thing ever made. You put a chainsaw in the hands of Nicholas Witchell and immediately he becomes as masculine as Schwarzenegger biceps. There's no way Prince Charles would have said 'I can't bear that man' if Nick had been brandishing a chainsaw rather than a microphone.

If you have a chainsaw in your hands, you are the most powerful person in the room. Politely ask Elon Musk to sell you a majority shareholding in Tesla and he'll tell you to go away. Ask him while revving a chainsaw and the company will be yours in a matter of moments.

The problem is that you wouldn't be able to rev a chainsaw because chainsaws are just about impossible to start. You yank away on the starter cord endlessly until your arm is weary, and then you remember that there's a safety catch that needs to be pressed before the motor will fire.

Unfortunately, there's a safety handle as well. So you need one arm to hold that down, one to hold the machine steady and one to pull the cord. There are more safety features on a modern chainsaw than there are on a Boeing Dreamliner.

Eventually you become so hot you have to take off your safety helmet because sweat is running into your eyes. And then you have to remove your gloves because you can't operate the safety switches, and then, after you've deployed your best swearing, it bursts into life.

And it is not what you were expecting. It is terrifying. You know that at any moment the chain will come off and sever your head. So you set off very gingerly towards the tree that needs surgery. And soon you will fall into a badger sett that you hadn't seen because of all the nettles.

It's thought that when a man is falling over, he's out

of control. Not when he has a running chainsaw in his hands he isn't. It's like falling into the sea when you're holding a mobile phone. You'll sacrifice yourself if necessary to keep it above water.

The Tubes were an American band that had very little success until frontman Fee Waybill, who sang under the name Quay Lewd, decided to appear on stage with a chainsaw. Everyone wanted to see that. Me especially. Which is why I was there, at the De Montfort Hall in Leicester, the night Fee fell off the platform shoes he'd made from old tomato juice cans. He could have put his arm down to save himself but that was needed to keep his chainsaw under control, so he didn't. As a result, he spiral-fractured the fibula in his right leg and that was the end of the tour.

Soon I reached the tree, my face lacerated by the brambles I'd fallen into, and started to cut. 'Zzzzz' went the motor, angrily. And then nothing. It had jammed. To unjam it, I needed to lift the branch slightly, which meant I was doing one-handed chainsawing. I'm not sure this is advised in the manual.

The other thing I learnt is that no matter how you approach a tree or branch, it always, always, always falls on your head.

After half an hour of sweaty manliness, I started to realize I didn't know why I was doing what I was doing.

Was I trying to get more sunlight on to the forest floor? Or less? Did I want to encourage the spread of the brambles or stop it? Woodland management sounds very important, but to be a manager you need to know what you're doing, and out there I started to feel like Matt Hancock.

I therefore called John Deere, which sent round easily the most fantastic machine ever made. It's a Nimitz-class destroyer of worlds. It's like a Star Wars battle cruiser has had sex with Edward Scissorhands, and what it does is breathtaking. The operator drives up to the tree and tells the machine what sort it is. After a quarter of a second, during which time it does maths to determine how many planks can be produced from the trunk, the tree is cut down, turned sideways and chopped up. Even the biggest tree in the wood is in strips in three seconds flat.

After an hour or two I went to a nearby hill for a picnic lunch and, looking back on the roof of the wood, it was like Godzilla was in there, having a temper tantrum.

After two days, my woods were ruined. Carpeted with sawdust, and with a whiff of diesel in the air, there were logs everywhere. Thousands of them. But as I stood there with my puny chainsaw, a tool I now realize is good only for chopping off a drug dealer's arms, it was explained to me by Godzilla's driver that I'd done the right thing.

The new environment will be good for deer, butter-flies, bees, flowers and the trees that remain. Plus, I can sell the timber I've created to the new green power station in Kent and make a profit of £3,000. Which will be just enough, come September, for the lunch I'm planning at the Colbert café.

Water, water everywhere.
Except where I need it

For such a famously wet country, Britain has always been notoriously useless at dealing with water.

We spent a fortune on dams and reservoirs when we thought industry would bring thousands of workers to the northern factories, and they opened at exactly the same time as all the factories shut and everyone moved down south.

Today there are 27 reservoirs in Derbyshire, 26 in Lancashire and 110 in Yorkshire, while in Hertfordshire there's one, in Kent there's two and in Hampshire there are none at all. And there never will be, because while it's easy to evict Albert Arkwright and his whippets from his hole in the ground in Heckmondwike, it's nigh-on impossible to get the Fotherington-Sorbets to move out of their pile in Odiham.

There was once a plan to fill giant polythene bags with water from those enormous northern reservoirs and, because fresh water is less dense than sea water, float them down the North Sea to the Thames estuary. But that would have been exciting and clever, so we got hosepipe bans instead. And we just accept that what comes out of the tap got there via the bladders of six other people.

I have a similar problem here on Diddly Squat Farm. There are ten springs that I know of and none of them is where I want it to be.

In the nineteenth century, a previous owner installed a pump, which was used to force water from one of the streams to a tank at the highest point of the farm. And then gravity carried the water from this, down a network of underground pipes, to the troughs he installed in every field. Brilliant. But the halfwit never made a map of where these pipes were. Or the tank.

A neighbour called Charlie, who may be mad, suggested that to find everything I should walk about with two coat hangers. In my mind water divining is like ley lines and horoscopes. It's nonsense. Two coat hangers will not react to the presence of water. Except for one thing. They do.

It was brilliant. They'd go berserk, I'd fire up the mini digger and, bugger me, right where the coat hangers had crossed, there was a pipe. I found them all over the place. The massive Victorian underground water engine was still there. So all I had to do was fill the tank at the top of the hill and it'd wheeze into life once more.

The old pump, made from leather and powered by

men with no teeth, had long gone, so I installed a new one, used a mole on the back of my Lamborghini tractor to dig a mile-long trench, and now the troughs are fully functional once more.

But it turns out I don't need them. They were installed before stewardship schemes and fertilizer and big tractors changed the way farming is done. Which means they're in fields full of nothing but marjoram and orchids and butterflies and ground-nesting birds. All of which can manage perfectly well without my subterranean water system.

What cannot manage are my vegetables. It is stupid to try to grow vegetables in this part of the country. The soil is wrong and because it's so high and exposed, it is below freezing most of the time. Summer here lasts from 2 July at 10 in the morning till just after lunch.

However, last year I ran a small potato experiment on a 2-acre plot and, contrary to the advice from absolutely everyone, they grew well. I ended up with about 40 tons of the damn things.

So when Covid-19 hit and there was panic-buying in the shops and borders were being closed, I had a wine-powered idea. As people would not be able to buy their vegetables from abroad, or even from Kent, if travel was banned, I'd grow some. Yes. I'd be the broad bean king of Chipping Norton. And the man you call late at night if you need an onion.

My land agent raised an eyebrow and suggested the idea was foolish. 'Ha,' I responded, full of the confidence you get after twenty years in Notting Hill. I pointed to a nearby field where we'd planned to grow spring barley and explained that, because of the rain last autumn, everyone would be planting the same thing. He agreed. And then I delivered the coup de grâce. Spring barley is used to make beer, and all the pubs are shut, so there'd be a glut of something no one wants anyway. 'Much better, then, to grow vegetables in it,' I declared triumphantly.

Planting the so-called 'sets' was tricky. I bought a machine from the Middle Ages, but that turned out to be useless. So Lisa and I did it by hand. By which I mean Lisa did it by hand. And having seen how much she was enjoying this, I decided to keep right on going.

Last year the dreaded flea beetle, which a man in Brussels says I'm not allowed to kill any more (rightly so, actually), destroyed a 10-acre field of oilseed rape. The field is therefore empty. And what's the point of that? Why not use it to grow pumpkins for Halloween and lavender for people's knicker drawers and sunflowers for . . . actually, I don't know what they're for.

Lisa was thrilled. I know this because she rolled her eyes, slammed the door and went for a long walk on her own to celebrate. I, meanwhile, ended up with a 14-acre vegetable patch, and as anyone with a window box knows, all I needed then was a regular supply of rain . . .

April was the fifth warmest since records began in 1884 and, while it went cold at the beginning of May, it didn't rain at all. I can't remember when it last rained here. The ground is parched, cracked. I'm living in a dust bowl.

Yes, tons of water are still pouring out of the springs and it's all being harnessed by my underwater engine, but it's then being fed to the wrong fields.

Desperate, I broke out the mole, got someone to fit it to the back of my tractor – I still can't do that – and created a new underground pipe to one field, which I must get round to marking on a map. I then attached this to some sprinklers, which have now seized up for no reason that I can see.

The other field, however, is on the far side of a small road and it seems I'm not allowed to dig a trench across that. So I bought a vacuum-operated slurry tank that sucks water from a stream and then sprays it over my vegetables. Unfortunately it also sprays it over everything else. Which means the field is now one part vegetable and nine parts thistle. I know now how Jean de Florette felt.

Last night, having marinated myself in more wine, I was looking into the possibility of using a hovercraft as a water dispenser. That's had to be shelved this morning, however, as the amount I've spent on my vegetable operation already means each broad bean will have to be sold for £17. And that's nearly as much as you'd pay at Daylesford.

There's only one solution as far as I can tell. I'm going to have to call Donald Sutherland and Kate Bush, and get the plans to that rain-making machine they made.

July

War with the wildlife

I'm not quite sure how I've managed this, but somehow I have reached the age of sixty without absorbing a single piece of information about trees. Literally nothing. I know more about Jane Austen, and all I know about her is that her Christian name is Jane, her surname is Austen and she wrote about a liberated young woman called Emmanuelle.

I must, occasionally, have been on a walk where someone started to talk about the trees we were seeing, but I guess I must have a filter in my head that turns tree talk into an eerie silence. I therefore cannot tell an oak from an ash or a spruce from a larch. They're all just green and brown and covered in bark. I only know what a Christmas tree is when it's covered in tinsel.

However, there are a hundred acres of woodland on my farm, and in the past nine months, since I decided to do farming for a living, I've had to try to learn something about how they work. This is tricky, because when I go into the gloom with a man who has no fingers – everyone in forestry has no fingers – he only ever gets to

'You see the thing about an oak is . . .' and the filter kicks in so after that I hear nothing at all. I had the same problem at school in chemistry lessons. 'Your lips move but I can't hear what you're saying.'

Despite all this, I have learnt some things. First, it is impossible for a tree to survive without man's help. If you plant one and then leave it alone, it will be eaten by a deer or a hare within a week.

To get round this, you must surround its spindly little trunk with a piece of plastic tubing that's designed to split when, after about ten hundred years, the trunk is wide enough to withstand attacks from Bambi and his overgrown rabbity mates.

At this point the grey squirrel will arrive and remove all of the bark to a height of about 2ft. This means the tree will become infected with something and die. Or it will grow more slowly than the other trees around it, which means it will be deprived of sunlight and die.

Eventually, and I genuinely don't know how it's possible, a few trees will grow to become big and strong, but this takes such a long time, you and your children will not live long enough to reap the rewards.

To get round this, I recently planted twenty trees – I don't know what they are; they're all brown and green – that were already 25ft high. Each one cost more than most hatchbacks. They arrived on a fleet of articulated lorries, with their roots encased in sacks, and were lowered into holes that had been made by a 21-ton digger. This was wilding, with extra diesel. And now it is my job to look after them.

It is a big responsibility. Twice a week I must pour exactly 25 litres of water into the roots of each tree via a tube that sticks out of the ground like an exhaust pipe. And another 25 litres around the trunk.

As there is no liquid refreshment in the field, it means I must first fill a tanker with a thousand litres of water and then spend two hours measuring it out and delivering it to precisely the right places. If I do not do this properly, the trees will die. So I am doing it properly. And, from what I can tell, the trees are dying.

This may or may not have something to do with a vast range of diseases that a tree can and will get. And the problem is going to get worse, because in the run-up to the last election, each of the main parties, and the Lib Dems, was promising vast tree-planting programmes in an effort to shut up Greta Thunberg.

We ended up with the Tories, who had said they would plant 30 million trees a year by 2025. That's 82,000 a day. Leaving aside the issue of who exactly would do all the planting, now we have left the EU, there's the bigger question of where they are going to find 30 million trees a year.

Abroad, is the obvious answer. But when you import a tree, it will arrive with bugs and fungi against which the native trees have no immunity. Dutch elm disease came from Canada. Ash dieback came from mainland Europe. So, to fulfil a political promise, we import one diseased tree from Finland and end up killing, according to recent estimates, 72 million trees that are already here.

There's another problem too. We will not be creating these 30 million trees. We will simply be moving them from their place of birth to Britain, where almost all of them will be killed by rabbits, deer, squirrels, disease, the growth ambitions of other trees . . . or me.

One of the things you learn when you become a countryman is that all real countrymen say the same thing when they walk into a wood. 'Hmm,' they chunter. 'This needs thinning.' That's what my keeper said to me. It's what my tractor driver and land agent said too.

So, in a single week I took 200 tons of timber from a 10-acre slab of woodland, and when I posted a picture on Instagram of the gigantic John Deere machine that I'd used, every single teenage girl who follows me – all four of them – came back with a stream of venom and anguish. I was worse than McDonald's. I was ruining their future and choking their grandparents. I was doing deforestation, and that's worse than racism.

Incredibly, however, it's almost impossible to tell that

any trees have been felled at all. The only difference is that now the forest floor is aglow with puddles of sunlight, which will stimulate all sorts of new growth.

In the past I've walked through that wood and it was ever such a dark and gloomy place. They could have filmed *The Blair Witch Project* in there. They probably did. But now there's new growth of nettles here and there, and for the first time in probably twenty years you occasionally trip over a hoop of bramble. By killing a bunch of trees, then, I've brought the wood back to life.

That's good for Bambi and the hares. It's good for the squirrels. It's good for the 250,000 bees I've just put in there, and it's good for all sorts of small flowers about which I know even less than I do about trees. It's also, according to my keeper, good for my shoot.

My exciting new hobby: beekeeping

Everyone likes Morgan Freeman. And now everyone likes him a little bit more because we learnt recently that he's turned his 124-acre Mississippi estate into a sanctuary for honeybees.

We have it in our heads that honeybees are important. And we are right. Being kind to bees is even more important than not throwing a plastic bottle into the sea, or not buying a Range Rover.

There's an impressive documentary called *The Serengeti Rules*, which explains that in each tiny ecosystem there is always one keystone species. You remove the barnacles from a rock pool in the Pacific Northwest and nothing happens. It's the same story if you remove any of the other things in there, except the starfish. If you get rid of them, pretty soon all you have left are mussels, because they are no longer prey.

The documentary-makers take us on an odyssey round the world, showing how this system works everywhere, until they end up on the Serengeti, where it turns out that every single thing owes its existence to the vast herds of wildebeest. Unless these are maintained, in huge numbers, nothing else can survive.

Which brings us back to the honeybee. If it becomes extinct, pretty soon you'll be killing your neighbour for a half-eaten tin of cat food and licking the moss in your cellar to stay alive.

Honeybees are responsible for about £20 billion worth of American crop production a year. The bee is the cornerstone of everything. It is the planet's keystone species. And for the past few years, in Europe, its numbers have been dropping at an alarming rate. Which is why I have stepped in and done a Morgan. I've decided that my farm should be bee-friendly.

That's why there are now 150-ft-wide strips of wild flowers growing in the middle of my spring barley fields. And it's why, three weeks ago, I took delivery of a quarter of a million bees.

They were delivered by a Ukrainian man called Victor, who said I must check on them every two days. So that's what I did. I stood about in the woods where I keep the hives, watching the bees whizzing hither and thither. And after a short while I realized I had no idea what I was looking for exactly. It was like checking my prostate. What's normal and what's not?

I read many books and was interested to learn that the bee that finds a large amount of nectar will return to her hive to perform a 'waggle' dance that lets the other bees know which direction they should fly to find it and how far away it is.

The bees calculate how much energy they'll use to cover that distance and therefore how much food they'll need for the return journey with the extra weight of all that pollen in the baskets on their back legs. This, of course, is just the lady bee. The gentleman bee does nothing. He sits in the hive all day with his mates, waiting for the queen to say she fancies a shag.

Life inside the hive is almost impossible to understand, but one thing's for sure: these guys have a society that makes the Austrians look sloppy and disorganized. They even keep the hive at precisely 35 °C on hot days by stationing themselves at key points in the structure and beating their wings. But not too much, as this dries the air, and they know honey can be made only if the moisture content in there is 17 per cent.

Ah yes, honey. The superfood to beat all superfoods. Hilariously, commercial beekeepers are told by the government's food standards people that they must put a 'best before' date on their jars. My Ukrainian friend Victor is getting some labels printed that say 'best before the end of days'.

But he'd still be wrong. Because honey never goes off, ever. They could have buried a rack of it with

Tutankhamun and it would still be as delicious today as it was then.

I read about bees solidly for a week and worked out they were definitely the inspiration for the Borg in *Star Trek*. So then I had to watch that. And afterwards I realized I still didn't know what I was supposed to be checking for on my visits to the hives.

So Victor came back to explain. Like Morgan Freeman, he doesn't wear any protective clothing. Morgan says that if you are on the bees' wavelength, they won't sting you. Victor says he is stung all the time and doesn't mind. I, meanwhile, was dressed up like Neil Armstrong.

We opened the first hive, and I'm going to be honest: I was staggered. Weak-kneed with amazement and joy. I'd read that in its entire six-week life a bee will only make a twelfth of a teaspoon of honey, and that to make 1lb it would have to travel 90,000 miles. Which is why, after just three weeks, I wasn't expecting much.

But just the top drawer of one hive was so heavy I could barely lift it. With Victor pumping smoke into the swarm to keep it calm – I don't understand that either – I pulled out a single frame from the drawer, or 'super', as it's called, and there was easily 2lb of honey in there. This meant 20lb in that drawer alone. And there were four drawers and five hives. So 400lb of honey. And they'd done all that in twenty-one days. As well as making all the honeycomb and producing enough wax to polish the floor of a Scottish castle.

I then learnt I had to examine the honeycomb for unusual developments. It turns out that this is impossible, mainly because one unusual development looks exactly the same as all the others. We were hunting for evidence that another queen was about to be born, which would cause half the colony to swarm, which is a polite way of saying 'bugger off'.

I couldn't even find the existing queen. Victor said she looked completely different from all the others, but when he located her, it turned out she wasn't completely different at all. A Volkswagen and a pencil are completely different. She was just a bit bigger and whoa . . .

As I examined her, one of her workers had noticed there was a 2-mm gap between the bottom of my space-suit and the top of my shoe. It was quite literally my Achilles' heel and she'd dived in there for a kamikaze attack.

A honeybee does not last long after she has stung something because, to get free, she has to pull her own arse off. So why had she stung me? I have no idea.

What I do know is that the scent of her poison sent the entire 250,000-strong army into a frenzy. As I hopped towards the car for cover, whimpering gently, my documentary cameraman was stung twice in the face and the director got one in the nostril. And now, three days later, I still can't really think straight because my foot hurts so much.

It's a price worth paying, though, partly because I'm now an eco-warrior, but also because since I started eating all the honey my bees made, I haven't had any hay fever at all.

I've also learnt how to stack a dishwasher properly and how to say 'no' to a second glass of wine. I may wash my car later and tidy my bedroom. Resistance is futile.

Attracting wildlife to my farm – with a JCB

It was the Victorians who decided that a garden should be neater and better organized than Jean Brodie's underwear drawer. But it's now been decided that 'tidying up' is very unvironmental and that nature should be left to its own devices. It's called 'wilding' and it's the new big thing.

When a tree falls over you leave it there for the beetles. When an animal dies you put a clothes peg on your nose and wait for the body to be devoured by birds. And you learn to love thistles and brambles and nettles to such an extent that you will sell your lawnmower for scrap. It is now considered to be the tool of the Luddite.

In farming, there are plans to ensure public money will be handed out only for public goods. In other words I won't get cash to help me grow food for humans; only for newts.

They tried something similar to this in Chile. Huge grants were made available to landowners for planting trees. So what the landowners did was to chop down ancient woods and forests, sell the timber and then take

taxpayer cash to replace it with a monoculture of industrial trees that have about as much eco-diversity as a fat
kid's lunchbox.

I was amazed by this because I didn't think there
were any forests left to chop down in that part of the
world. When I went to Tierra del Fuego a few years ago,
I was genuinely staggered by the destruction. It was as
though there'd been an atom bomb and a hurricane at
the same time. For mile after mile every single tree was
on its side, dead.

And all of this – contrary to what you will read on eco
websites – was as a direct result of a 1946 Argentine
wilding project that brought ten pairs of beavers to the
area from Canada. There are now more than 100,000
animals, and recent research has revealed they've been
responsible for the biggest landscaping alteration in sub-
Antarctic forests in the past 10,000 years.

As the scale of the problem became clear, a Chilean
environmentalist, Felipe Guerra Díaz, hoped the beavers would be stopped at the border. Fat chance. 'They
don't recognize borders,' he said. 'In fact, they eat the
border fence.'

We've seen this sort of thing before. When America
created Yellowstone National Park, experts reckoned that
visitors would not want to be eaten. So it was decided to
get rid of the wolves. But with no wolves, elks flourished

to such an extent that all the aspen and willow trees were eaten, and the effects of that wiped out countless other species, including the beaver.

It's just a fact that governments never know what they're doing: I mean, that track-and-trace app — how did anyone ever think it was going to work?

But all of us quite fancy the idea of bringing a bit of the wild back into our lives, which is why I decided to create a wetland area on the farm (I didn't take a grant, before you all start growling). In my mind, I'd clog up one of the streams with a small dam and then allow bulrushes and water irises to sprout from the resulting bog. Simple.

Obviously, though, the government has strong views on what you can and can't do with water that flows through, or bubbles up on, your farm. You can't divert a watercourse any more than you can divert a footpath. And while you are allowed to use a bit of the water for irrigation, the amount is strictly controlled.

You also need permission from a man in a Vauxhall — all government officials have Vauxhalls — if you want to change the bank, dredge, build a culvert, change a mooring, build a dam, create a weir, take a fish or park a boat.

I tried to argue that my watercourse was not a river, or even a stream. It doesn't even really qualify as a beck, but it didn't matter: a man came — in a Vauxhall — found a small pile of ordure and said he thought the area was home to some water voles. And as they're aquatic bats, everything had to stop.

Trap cameras, however, revealed that the vole was, in fact, a mouse, so I was able to start my nature wilding project. I therefore fired up the 21-ton JCB and pretty soon the whole area was shimmering in a reassuring haze of diesel exhaust and you had to shout to make yourself heard over the sound of internal combustion. Plus, the digger's tracks had made the whole area look like Sam Mendes was filming a sequel to *1917 – 1918*, perhaps.

I was very happy, but as I'm not as good at operating a digger as I like to think I am, the hole I'd dug was becoming way deeper than I'd imagined. I just wasn't able to operate the bucket accurately, so every time I wanted to scrape a bit of earth into a level bit of bog, I ended up taking away an extra 4ft from the lithosphere.

The plan was to slow the water down, so that the wetland could be a rich natural playground for biosustainable, environmental, fair-trade, nuclear-free diversity in the community. It'd be a place to take the knee and clap the NHS at the same time. And it would also help prevent flooding problems downstream.

However, the hole I'd dug was deeper than an Australian uranium mine, and that meant the earth bank that was being erected from the spoil was larger than the Three Gorges Dam in China. It certainly wasn't what the man in the Vauxhall had given me permission to create.

Finally, though, I ended up with a clay-lined pond maybe 50 yards long and 10 wide, and one end of it was shallow enough to become home for some reeds and bulrushes. I even transplanted some stripy weeds I'd found in another pond, and they seem to have taken well.

I'm using solar power to pump water on to the banks, so the wild-flower seeds can germinate, and I've bought 250 trout. That has brought cormorants from the coast and a family of otters from God knows where. There's a heron too, but herons are the most bone-idle creatures on God's earth, so it just spends its time looking at the fish, thinking, 'If this bank weren't quite so steep and bumpy, I'd go in and get one.'

Does it feel 'wild'? No, not really. Especially now there's an electric fence to keep the otters out, nets to keep the cormorants at bay and a pontoon to stop the trout getting sunburnt. Yes, I know, I didn't believe that either, but apparently it's true.

However, while I was down there last night feeding the fish some entirely unnatural food that looks like rabbit droppings but smells like Bhopal, I noticed a dragonfly hovering above the water. It was a whopper, and it had a body finished in a vivid metallic turquoise. I was thrilled, because it wouldn't have been there were it not for my labours. And neither would the kingfisher that darted out of the reeds and ate it.

August

Shearing Jean-Claude Van Lamb

I decided recently that my sheep are like woolly teenage boys. They take absurd risks and feign a lack of interest in everything, while deliberately being obstructive, stubborn, rude and prone to acts of eye-rolling vandalism. I understand this. Between the ages of fifteen and seventeen I could not walk past a fire extinguisher without setting it off. That's a very sheep thing to do.

A week ago I put them in a field full of succulent grass, with far-reaching views of the Cotswolds and the Chilterns beyond. It was sheep paradise in there, but every single day, without fail, all of them would walk straight through the electric fence and into a neighbouring field full of not-at-all succulent spring barley.

There are two reasons why they do this. They know that, thanks to the weird weather we've had in the past nine months, the profit margins on barley are extremely tight, so if they eat some of it they will cause me financial hardship. They also know that barley, if eaten in sufficient quantities, will kill them, and that, as always, is their main goal in life. To die, horribly. If sheep could operate machinery, they'd all have very powerful Japanese motorcycles.

Last week they had to be sheared. Now, you may think I'd do this for my benefit and sell the wool for profit. Ha. How wrong you are.

The first problem is that the Aussie and Kiwi chaps who roll around the world like a wave, chasing the shearing seasons, are marooned this year by Covid-19, so the price for skilled labour has gone up. Reckon on about £1.50 per sheep. And how much do you get for one fleece these days?

Well, as they're not being exported to China and they're not being used to make carpets and rugs here at home, the price I'd get for the wool from one North Country Mule is ... drum roll ... 30p. So I'm losing £1.20 per sheep on the deal. Now you can see why I call my farm Diddly Squat.

The only reason, then, I had to get shearing is because a shaved sheep is less likely to be eaten by maggots and is more comfortable in the summer sunshine.

You'd imagine they'd be grateful, but no. The first sheep took one look at the salon I'd made and charged at top speed into a bramble bush, where it writhed about until it was stuck fast. Freeing it meant lacerating my arms and legs so extensively that I was in danger of

going into hypovolaemic shock. And by the time I'd finished all the other sheep had wandered through the
electric fence again and into the barley field.

I had two shearers on hand, but the idea was that I'd
get stuck in as well. So after we finally got all the sheep
into the pen, I set to work.

Job one is to capture a sheep, turn it upside down
and drag it on its buttocks into the shearing area. This is
not possible because sheep do not want to be turned
upside down and dragged about on their buttocks. And
they are very strong. Imagine turning Jean-Claude Van
Damme upside down and dragging him into a salon to
cut off his mullet. You could explain as much as you
like that he'd be better off without it, but you wouldn't
get anywhere. I was knocked over, kicked and then
knocked over again. The professionals advised me I
should grab the sheep under its chin and fold it in half
before turning it over, so I tried that, and I was kicked
before falling over again.

Some of the other sheep decided at this point they
didn't like sharing the pen with a seemingly violent fat
man who kept falling over, so they jumped out. This
amazed me because even though they had no run-up
space at all, they managed to clear a fence more than 3 ft
tall. Honestly, sheep have better VTOL properties than
a Harrier jump jet.

Eventually, though, I managed to get one of them in the right position and in the right place, and then I was told to stick one of its back legs up my butt crack and reach for the shears.

These were the most fearsome things I'd ever seen. Imagine the inside of a gearbox on the sort of circular saw they use in Canadian logging yards. And now imagine wielding that while trying to cut off Mr Van Damme's hair. There could be only two outcomes. The sheep would die. Or I would.

'Go on,' said the onlookers. But that seemed foolish. Bomb disposal officers don't urge onlookers to give it a bash. A surgeon doing an eye operation on a young girl doesn't ask the father if he'd like to wield the scalpel. So I put down the oscillating, out-of-control lumberjack's gearbox and went on strike.

The professionals got to work, taking less than two minutes to give each sheep a buzz cut, and I, meanwhile, was given the lesser job of rolling up the fleeces and putting them in the woolsack. Though when I say 'lesser', what I mean is 'revolting', because before the fleeces can be rolled, you have to separate the dingleberries from the wool. By hand. You have to get down on your knees and tear chunks of dried faeces from the hair. And if you tear too much hair away, you are shouted at, because it's worth something. It is too. About 1p at current prices.

When we'd finished, we had the job of herding the 75 shaved sheep and 145 of their still woolly offspring into a new field on the other side of the farm. Two dogs were used to keep them in line and I was put in front of the flock with strict instructions not to let the sheep overtake me.

I'd like to say I got a hundred yards, but it was actually about 25 before they tore past and ran into a field of barley for a lethal afternoon snack. And as I stood there, trying to rub the pollen out of my eyes while country folk told me that rubbing my eyes is the worst thing you can do, I realized that I am not, and probably never will be, much of a sheep farmer.

However, there is a happy ending. I'm building a house and I noticed the other day that the builders had started to fill the wall cavities with glass fibre.

Wool does a better job. It has a low thermal conductivity rating – which means it's good at stopping heat escaping – it's flame-retardant, it can absorb water vapour better than almost anything else and, because it's a protein, it can remove unpleasant odours through a process called chemisorption.

So instead of selling my wool for the square root of

bugger all, I'm going to use it in the cavities of my house. This means I won't need to buy the expensive man-made alternative. So while my wool won't make me a penny, it'll save me thousands. And get me a letter of commendation from Greta Thunberg.

What's happened to my crops?

When I was a small boy I went to a Church of England primary school in a mining village outside Doncaster, and every September we were frogmarched into the local church to thank God for the harvest. But as I sat there, staring at the altar, which was festooned with trugs full of marrows and ears of corn, I always used to think, 'God had nothing to do with any of that. It was all produced by Mr Turnbull,' who was the local farmer.

This year God definitely didn't have anything to do with it. In fact, he's done his best over the past nine months to make sure there was no harvest at all. He gave us the wettest autumn since 2000, the wettest February on record, the driest May on record and then, for good measure, the coldest July since 1988. He's fried my crops, frozen them, drowned them and then drowned them again.

So if we are going to thank anyone for the harvest in 2020, might I suggest we sink to our knees and give praise to the giant agrichemical group Monsanto, whose weed-killing glyphosate invention has enabled me to keep my head above water. And while we are at it, let's not forget Syngenta, Bayer and BASF, which make the nitrates and the pesticides. It's thanks to their efforts, not God's, that you're going to have bread on your table this year.

That said, as harvest approached, I was warned by my land agent that despite the heroic efforts of the chem-science boys, the crop yield would be poor. However, on my daily walks, emboldened by the gift of innocence, I would gaze upon my barley and my wheat and my oil-seed rape and think, 'Well, it all looks fine to me.'

And it continued to look fine until, one day, Kaleb, my tractor driver, went out and killed all the rape with a potent cocktail of weedkillers. Apparently this is necessary. Rape has to be dead before you can harvest it. Who knew?

Another thing I didn't know is that you can't harvest rape if the moisture content of the tiny seeds is greater than 9 per cent or less than 6 per cent. So if it's been raining, or not raining, or if it's been sunny, or not sunny, you have to wait. Too dry and the crop will be rejected by the buyer because it will be impossible to extract the oil. Too wet and you must spend a fortune drying it out.

There were more complications too. Earlier in the year, when my land agent asked where I was going to store all the harvested crops, I said, 'In a bucket.' Judging from the incredulous look on his face this was the wrong answer, so I said, tentatively, 'In the bath?' This was also wrong. You need a barn – so I built one. It was huge, the kind of thing they use to make airships, but when he arrived to inspect it, he said it would be big enough only to store the rape.

This meant that before harvesting could begin I'd need to organize a fleet of trucks to take away the barley. I'd also need to rent a combine harvester and, although this is not a common problem in farming, book a film crew to cover the event for the Amazon television show I'm making.

This meant I needed an accurate weather forecast, so on the Sunday evening I sat down to watch the BBC's *Countryfile* show. And in the 'what's the weather got in store for the week ahead' segment, the forecaster said high pressure was on its way and we could expect clear skies, temperatures in the mid-twenties and light winds.

The next day it was cold and wet, and I was furious. To you, inaccurate weather forecasts don't matter. The worst consequence is you have to abandon the barbecue you'd planned and move inside, but to a farmer they are critical, so I have a plea to the Beeb's weather people: if you don't know – and at the moment you don't, because the transatlantic pilots on whom you rely for information are all at home learning how to make sourdough bread – admit it.

There's no shame in that. I don't know lots of things. I don't know the boiling point of steel or how to make a roux or who painted *Belshazzar's Feast*, and I'm big enough to admit to these failings. You should be too. Especially on a farming show.

The next day I called my neighbouring farmers to say I was going to have a coronary, and they all had the same piece of advice. I had to accept whatever happens, because that's farming. They also said I had to be patient, which is not possible. I can't be patient. It's not in my DNA. It's a bit like telling Nicholas Witchell he has to be a Moroccan cage fighter.

But after a while I'd calmed down to the point where my heart was beating only a million times a minute and then, contrary to what we'd been told in that morning's forecast, we got one of those grey days when it feels like God's put the whole country in a Tupperware box. So we placed a handful of rape seeds in my £500 moisture-o-meter and, after a heart-stopping pause, got the result: 7.2 per cent. Perfection. We could begin.

Except we couldn't, because the trailer I'd bought had hydraulic brakes and my Lamborghini tractor has a system operated by air. Stopping was tricky with nothing in the back, but when I had 12 tons of crop back there, I'd be very much on the wrong side of both the law and the road. So I rented a trailer with air brakes, waited for another unforecasted weather front to pass and then everything began. Again.

And five minutes later I received the news that a whole year of properly hard graft had been a complete waste of time. The combine has a computer that can tell how big the crop is. In a good year I could expect maybe 1.5 tons per acre, but the initial figure said I was getting just 800kg. I'd been prepared for bad news, but not that bad. And it was about to get worse.

My job was to wait at the side of the field for a flashing light on the roof of the combine to begin twirling. This would announce that its hopper was 80 per cent full. I'd then rush over and drive alongside the mighty harvester while a fan blew the seeds up a tube and into my trailer.

All I had to do was drive at the same speed as the combine. Simple, yes? No, actually. You need to look where you're going, so you can drive in a dead straight line, but you also need to look backwards to make sure the seeds are going into the right part of the trailer. To do it properly you need eyes on the side of your head. Basically, you need to be a pigeon, and I'm not, which is why, in the first few minutes, most of the miserable quantities that were being harvested were being immediately onanised back into the ground.

Despite all this, we worked into the night. The next

day, when the dew had evaporated and the moisture content was back in the window, we started again until, that evening, my new R101 barn was filled with a big black dune of what was 50 per cent rape and 50 per cent earwigs. Apparently they will be filtered out before you use the oil to make your supper. I hope so.

And then came the real heartbreak. Brexit has buggered my barley. Because no one knows what will happen to international markets after the end of the year, there's a mad scramble to sell barley to Europe before the door is closed. So not only was the yield down by 20 per cent, thanks to God, but, thanks to the coffin-dodgers who wanted to 'take back control', prices are down too.

Which is why I'm writing this in Manchester. It's been said that, to make ends meet, farmers must diversify, and I have: later on I shall be recording the first in a new series of *Who Wants to Be a Millionaire?* I think I may sit in the wrong chair and get the contestant to ask me the questions.

Only Sting and the super-rich can save the countryside

There is a sound argument for handing all Britain's countryside and farmland to rock stars and bankers. And then giving them massive tax breaks to help them to run it properly. I'm being serious, because who else would you trust with such a big job?

Not the government, obviously. The government can't be trusted to do anything properly. I mean, all it had to do when it saw the pandemic coming was buy some aprons and some gloves, but somehow it managed to make a mess of it.

It was the same story in Iraq. Our troops desperately needed body armour and bullets and transport made from something stronger than Kleenex, and what they were sent instead was 6,000 pairs of chef's trousers. So there is no way in hell you'd let the government grow food or manage a hillside in Yorkshire.

There was talk by the Labour Party, when Corbyn was in charge, that land should be confiscated from the rich and given to the poor. But that wouldn't work either, because what poor people do when they have a bit of land is use it to store their rusting old cookers and vans.

I'm not sure farmers are the answer either, because farmers need the land to be profitable. So when they look at an agreeable view full of dry-stone walls and bustling hedgerows and ancient woodland, they don't think, 'Wow, this is pretty.' They think, 'Hmmm. I must fire up the bulldozer.'

I get that. If you invest a million pounds in shares, you expect some returns, and it's the same story if you invest a million pounds in land. So you're going to squeeze as much as possible from every square inch. Plus, of course, people want cheap food.

But there is a problem with all that. Since the Thirties, Britain has lost 97 per cent of its wild-flower meadows. This is because wild flowers look good on a postcard but terrible on a profit and loss account. There's simply no money in cornflowers and dog daisies, so any farmer is going to replace them with wheat and barley and oil-seed rape.

My farm, however, has always been owned by rich people who wanted land for shooting or hunting or avoiding tax. They no doubt saw the act of 'farming' as a bit grubby. A bit trade. This means the wild-flower meadows still exist. They were never ploughed up.

I hate flowers. They bore me. But they do not bore

everyone. They certainly didn't bore a man who came to my farm earlier this month. He was literally jumping up and down saying that what I have is pretty much seen nowhere else in the whole of the United Kingdom.

Apparently wild flowers don't do terribly well in Mrs Miggins's cottage garden. They need space to thrive, and that's what I have. Six big fields. A total area of perhaps 200 acres, which is chock-full of small scabious, kidney vetch, green-winged orchids, yellow rattle and various other things that sound as if they've escaped from a Victorian book of diseases.

My new friend said it was a staggering natural resource. Which is why I was a bit surprised when he climbed into his Land Rover, attached a small vacuum cleaner to the back and drove up and down the fields, sucking seeds into a hopper.

He then laid out what he'd collected on a giant tarpaulin and invited me to take a closer look. Which is a bit like asking someone with a severe nut allergy to apply for a job with KP. My hay fever went berserk, but in between sneezes, and through streaming eyes, I could see that the haul was moving. It was alive with insects.

I must admit I was excited, because wild-flower seeds at my local garden centre cost about £60 for 100 grams. And I had about half a tonne of the damn things. I therefore rushed home and immediately ordered a Bentley.

However, it turns out that my seeds will not be sold. They will be given away, which is not a phrase the Yorkshire part of my brain likes, or even understands. Apparently, it's all part of a conservation scheme funded by the Esmée Fairbairn Foundation, which was set up – surprise, surprise – by a wealthy investor in the early Sixties. The idea is that any of my neighbouring farmers who are suffering from eco-guilt can get hold of seeds that are genetically suited to this specific part of the world.

And, I'm told, the system works for me too because, thanks to my generosity, the government may be more inclined, down the line, to look favourably on my requests for public money.

And there's the problem. I don't want jam tomorrow. I want a Bentley today. I want to monetize my marjoram and fleece the bejesus out of my fairy flax. And I want to sell orchid seeds in my shop, like cocaine, only with a higher price tag. This is because I'm not rich enough to be a landowner.

Sting is. He has 800 acres in Wiltshire on which he farms pigs and hens, but not in the way that normal farmers do farming. It's all tantric and aesthetic, because he's not really interested in using the land to make money. He doesn't need to, because every time someone uses 'Roxanne' in an advert for panty liners, he can buy another organic jet.

If you think that the land should be given back to nature and managed sensitively, with insects as the No. 1 priority – and I'm beginning to think this way – then it is imperative that Sting is encouraged to buy as many neighbouring farms as possible. I may buy *Outlandos d'Amour* again this afternoon to help him out, and you should too.

Sting's not alone, either. For a long time, Ian Anderson from Jethro Tull owned a salmon farm on the Strathaird peninsula, on the Isle of Skye. Steve Winwood, of the Spencer Davis Group, Traffic and Blind Faith, has a chunk of land in Gloucestershire, and Alex James of Blur grows quite the nicest vegetables you've ever tasted on his farm just down the road from me in Chipping Norton. All of them should be given more farmland to play with as soon as possible.

Thirty years ago, if you drove through the countryside on a summer's day, you wouldn't get five miles before your windscreen was spattered with a million dead insects. Not any more. You won't hit one, because there are hardly any left. And that's bad, because without insects there will be no life on earth of any kind.

We need them, and we need hedgerows and ancient woodlands and areas that are boggy and sad. Almost everyone seems to be in agreement on this: we need to put Mother Nature back in the driving seat. And I'm sorry but the only way this is going to be possible is if we hand over control of our land to the super-rich.

The government should therefore consider this suggestion seriously: that we turn Britain's green and pleasant bits into the new Monaco, a tax haven for people who have both the time and the money to do the right thing.

AUTUMN

September

Why every farmer needs a flamethrower

When you look at a Royal Navy F-35 Lightning landing vertically on an aircraft carrier, you can't help but marvel at the technology that makes such a manoeuvre possible. But I can assure you that when it comes to engineering complexity, a combine harvester makes the fighter jet look like a toaster.

I've peered inside a combine and I've studied many YouTube videos on how they work, but I still reel in awe when I see one cut through a field of wheat and then, in a blur of noise and vibration, somehow manage to separate the grains from everything else.

Explaining how this happens is like explaining witchcraft, but in simple terms you wouldn't want to fall into any of it. There are a lot of knives, a lot of fans, a lot of jiggling and a lot of sensors that tell the driver exactly what's going on.

He knows, for example, precisely how many grams of grain he's getting from each acre of a field, and how wet it is, and, to make sure he can concentrate on that day's Test match, he doesn't even have to do steering. That's all done from space. An F-35 Lightning, meanwhile, has two wings, a joystick and no Test cricket on the stereo.

I will grant you, however, that the jet fighter does look better than a combine. Other things that look better than a combine include most wheelbarrows, your chest freezer, the marabou stork, the Chrysler PT Cruiser and Kim Jong-un's hair.

This is an issue that runs throughout farming. There's been a small amount of effort made in recent years to give tractors a bit of snazziness, but everything else is designed to do a job and put on sale. Even vacuum cleaners and lawnmowers are styled these days, but agricultural equipment? No.

It's true that farmers like kit. They like a new toy and a new gadget, but they also don't want to throw away cash on superficial flimflam. They don't, for instance, want their cultivator to have a spoiler or alloy wheels, because they know farming equipment is regularly bashed into gateposts and dragged through mud, and that it often lives outside, in all weathers, becoming rusty and stiff, until one day it has to be welded back together again by a cross-eyed sixteen-year-old halfwit.

All of which brings me on to my seed drill. This is a machine that makes even the combine look as simple as a wooden bench, because it can deliver seeds into a field at precisely the right intervals and at precisely the right depth. It even turns off two of the drills every so often to create a series of 'tramlines' of unplanted earth, which the sprayer can drive along when the crop starts to grow. What's more, it's all controlled by a laptop in the cab, so it's far more sophisticated than your iPhone, and yet it looks like the sort of thing you'd find in a Bangladeshi scrapyard.

If Cyrus McCormick, the nineteenth-century inventor of the reaper – widely regarded as the first bit of mechanized farming equipment – came back today and saw what was parked in my yard, he'd assume humankind hadn't advanced by even an inch.

I have a mole that allows me to feed water pipes into the ground and then bury them as I drive along. It's simple and clever and about as attractive as a genital wart. I also have a topper, which works brilliantly but has exactly the same styling you get on a heating-oil tank.

I've seen footage of a flamethrower being towed behind a tractor and I immediately wanted one, because who wouldn't want a flamethrower attachment? I'd have one on my ironing board if I could. But this one looked as if it had been made out of scaffolding poles and cow bells.

Part of the problem, I suspect, is that a lot of farming equipment has been designed by farmers themselves. And you only have to look at a farmer's three-piece suite to know that aesthetics rarely feature in his list of 'important things'. He just wants somewhere to sit down, so that's what he buys. This is a man who thinks uPVC windows make sense.

The farmer doesn't care what you think of his shoes, which is why he wears big plastic boots with metal toe-caps. He wears overalls that make him look fat. He cuts his hair by dipping it annually into his combine harvester and he continues to wear an oily tie that he found, fifteen years ago, holding the leaf springs on his trailer together.

He even manages to make his quad bike look dull and practical. Elsewhere in the world quad bikes are purple and have stripes, but here in England's green bits they have narrow wheels and mittens on the handlebars and they look like the sort of thing that was used to mow the grass at Biggin Hill in the summer of 1940.

There is, however, an exception to all this. The JCB telehandler, a machine with a telescopic arm that can lift and move heavy loads. I managed to go for fifty-nine years without one in my life, and now I have no idea how. Yesterday I used it to transport crates of empty bottles to my new water-bottling plant and, even though it's me we're talking about, and I'm neither practical nor careful, I didn't drop one. It's as gentle, then, as an eye surgeon's bedside manner.

This morning I used it to take stone to the new dam

I still haven't finished, and later I'm expecting a delivery of rape that will need to be stored in the barn. This evening, after I've used it to shovel the wheat into a neat pile, load some barley on to a truck and fetch some logs, I shall go to the pub in it, then tomorrow I will put a pallet on the forks and raise a child high into the apple trees to collect the hard-to-reach fruit. Apparently you're not supposed to use it for this purpose, but I can't see why.

Like all farming equipment it is extremely clever and very strong. You know when you pick up a matchstick? Well, that's how the telehandler feels when you use it to pick up a rock the size of a skyscraper. You're sitting in the air-conditioned cab, making tiny Neil Armstrong thruster movements on the joystick, and with no discernible effort at all it is rearranging geology.

It took God six days to make the world and we're all supposed to be impressed by that. But a telehandler could smash it up again in two. The machine I'm using can lift three tons 50ft in the air and not feel even remotely troubled by it.

And here's the best bit. Unlike anything else in the farmer's barn, it's as cool as the kit they used on *Thunderbirds*. It looks as if it were designed by someone who has a Poggenpohl kitchen and furniture made in Denmark. It is the best of both worlds, then: something you want and something you need. And now, because I'm not a proper farmer yet, I'm tempted to fit it with flame-throwers modelled on the guns in *Aliens*, and maybe some lasers.

How I turned around my little farm shop of horrors

This year I will produce probably 300 tons of wheat, 700 tons of barley and 250 tons of oilseed rape. And I'll almost certainly make a loss on every single ounce.

The cost of preparing the soil and buying the seeds and planting them and buying the fertilizer and pesticides and fungicides and then hiring a combine harvester to collect the crops and then storing them in a big fan-heater to dry them is greater than the market will stand. Put simply, it costs more to make your bread than you're prepared to pay.

It's the same story with my trout and my hens, and it's especially the same story with my lambs. Right now I'm faced with a choice of selling them as they are or paying for some supplementary feed and selling them later, all fattened up. I've done the maths and either way I lose exactly the same amount of money.

This is why so many farmers are opening farm shops. They make sense, because the food goes from farm to fork through no middlemen at all. And you, the customer, can buy your lunch from the field in which it was grown. People seem to like that.

Naturally I leapt on to the bandwagon and, earlier this year, built a small unheated stone barn with a view to selling whatever happened to be in season at the time. Lovely. Very Italian. But when the shop opened, the only thing that was in season was the potatoes.

Then Covid-19 came along, which meant I had to close the doors, and when I opened them again, in late June, all the spuds had gone to seed. It was a disaster.

And virtually nothing else was ready to be picked. The apples were tiny sour bullets, the wasabi was string, the vegetables were barely out of the ground and my water-bottling plant was still just a pipe dream. Actually, scratch that: it was only a pipe.

So I had to buy stock from other farmers, who also have farm shops, which may turn out to be bad business. I also had to contact local wholesalers, which is why, on the day of the grand reopening, we had a basket full of what I'm selling as Cotswold pineapples. And two avocados, which Lisa ate because she's a girl and no girl can walk past an avocado without eating it.

I desperately want to sell what I'm growing but even when it's all ready, I'm not sure it's stuff people want to buy. I mean, who wants an ear of corn? Or a bit of rape? It'd be like buying a cog when what you need is a car. In order, then, to turn these raw materials into what you'd recognize as food, I'd need to build some kind of factory.

I did the sums on that and the cost of the factory, plus the cost of fitting it out with ovens and stainless-steel drums with dials on the side – all food factories have those – and staff in crisp white uniforms and masks, means each loaf of bread would have to cost approximately £15,000. My beer, meanwhile, would be £36,000 a pint and my trout pâté £400 a gram.

What I could sell, however, at a reasonable cost and straight away, was the honey from my 250,000 bees. So I went to the hives, collected the supers, spun the trays and after two days had enough to fill the boot of my Range Rover. This was excellent, apart from the fact that all of it sold in less than two hours and there won't be any more for weeks.

It's much the same story with the sausages. My tractor driver, Kaleb, keeps some pigs and from time to time turns them into bangers that are quite simply the best thing I've ever put in my mouth. I sold four to one couple who came back the next day to get some more, and were noisily disappointed to find I'd sold out and wouldn't be getting any more until next year.

That's the thing about selling seasonal products on a small scale. It is extremely inconvenient for the customer. Today, for instance, I have rhubarb, but you'd better hurry if you fancy some because I only have eight stalks and there's no chance of a top-up until May.

I do have plenty of chard, though, which to begin with wasn't selling at all. It turns out that people round here (including me) don't know what chard is, so I'm now calling it spinach and it's flying off the shelves. That's probably illegal.

There are other issues. I accidentally built the barn about as far away from a power supply as it's possible to get, so I've had to run an extension flex to a nearby caravan site. The water comes from there too. There are no lavatories and one of the fridges I bought – for £1, so I can't really complain – sounds like a Foxbat jet.

On the upside I've installed a milk-dispensing machine. It's brilliant. You bring your own bottle, put it in a little glass box, where it's cleaned, and then, after you've put a pound in the slot, you get a litre of extremely delicious chilled cow juice. Milk 24/7 sounds great, and it is, but as the machine cost £6,000, I'm going to have to sell 10,500 pints to pay for it. That's a prospect even Asda would find daunting. And it doesn't even include the cost of the milk.

The funny thing is, though, that since the shop opened, all sorts of people have come along to ask if I can sell the stuff they're making in their sheds and on their allotments and in their back rooms, and, bit by bit, the shelves have been filling up with all sorts of tasty comestibles.

There's a woman who makes cakes and another who does sausage rolls and a man who came round with the best tomato juice I've ever tasted. There's even a rock star in the area who does cheese and, what's more, as the after-effects of Covid-19 begin to bite and more people lose their dreary office jobs, I suspect I'll have an even greater local pool of artisans to draw on.

But for now trade is brisk, the feedback is good and I've even taken on a shop girl to help out. I did the maths last night and I reckon that, with a fair wind, I'll lose only about £500 a month on the project.

It'll probably be more, though, because when the council gets back to work properly, it's bound to drop round and say I'm breaching some kind of bylaw, or that I must install lavatories, or that the roof is the wrong shape.

Or it could be that people will realize that while a farm shop is a great place to buy Scotch eggs during a pandemic, it's probably better in normal times to stick with a supermarket, where you can also get batteries and shaving foam and wine. And avocados, obviously.

I hope not, because while farm shops are not a solution to the desperate financial problems facing farmers, they do mean that the poor chap will be going bust a bit less quickly than if he sold his stuff to Lidl. And they are also a genuinely nice place to buy bloody good food.

October

Eau no, my pipe dream's sprung a leak

Here's something juicy to get your head round this morning: 97.2 per cent of the water on Earth is in the oceans, and a little more than 2 per cent is stored as ice in glaciers and at the poles. Whereas just 0.023 per cent is to be found in our lakes, inland seas, rivers, soil and atmosphere.

Happily, about thirty times more than that is stored as ground water beneath our feet. If you go out today and dig a hole in your back garden, you will eventually find it, whether you live in Alice Springs, St Petersburg, Montevideo or Hemel Hempstead.

Even the Sahara Desert is floating on a vast underground 'lake'. A study has suggested it covers a massive area beneath Libya, Chad and Algeria and could be 250ft deep.

So there we are, then. Problem solved. We can bathe and shower until we glisten with a pinky, wholesome goodness. We can water our gardens until our plants are giddy with the refreshing zestiness of it all, and the entire population of the planet can hydrate itself until we all look like some kind of hosepipe-based accident in a Tom and Jerry cartoon.

What's more, we have it in our minds that the water deep below the surface of the Earth fell as rain perhaps three million years ago and has spent all that time absorbing enriching minerals from the rocks it has passed through, so that it will make our brains big and our colons clean.

That's certainly what I thought when I sank a borehole on the farm earlier this year. The drill went down 300ft. A pump was inserted. And out came . . . well, it's tricky to say what exactly.

It looked like water and it smelt like water, which is to say it smelt of nothing at all. But after just two months the irrigation system in the fields jammed up, all the crops were covered in a weird white residue and at home the dishwasher, washing machine and shower all ground to a halt.

Tests revealed that the borehole was delivering a curious and possibly lethal cocktail of manganese, sodium and sulphates. The levels were so far beyond legal limits that if anyone even stepped in a puddle of it, they'd immediately grow two heads.

The problem is that there's only a finite amount of water on Earth. What we have now is what the dinosaurs lived on. It's what the amoebae climbed out of, after they had grown legs. It's what cooled the volcanoes back when everything was hot and messy, and, if you're that way inclined, it's what God used to water the apple tree in the Garden of Eden.

It's said that if you drink tap water in London, it will have passed through at least six other people before it got to you, but that's nonsense. It will have passed through many more than that, and a few dogs, and the odd woolly mammoth, and even a few brontosauruses.

It will also have passed through rock that doesn't necessarily have the rejuvenating properties of French limestone or Alpine granite. Rock such as we have here in the Cotswolds. Dirty rock. Diseased rock.

To solve the problem I was told to spend several billion pounds on a reverse osmosis system that takes everything out of the water, apart from the hydrogen and the oxygen. This would allow me to put back in what I wanted. A hint of David Ginola with a touch of Timotei waterfall, and perhaps a high note of wood-smoked apple blossom.

I didn't fancy that, so instead I turned my attention to the springs that bubble up all over the farm. I have no idea why this spring water might be different from the water I obtained through the borehole, and neither does anyone else. I think it's fair to say we know more about the surface of Mars than we do about what's happening deep beneath our feet.

But for some reason it is different. Some of it is full of E. coli, some of it heavy with nitrates and some of it a blizzard of faecal matter. But the test results from one spring came back with a clean bill of health. It was perfect.

Further investigations revealed that the nearby village used to live on it until one night in 1972 when the water board switched it over to the mains. There was a near riot. Questions were asked in the House. Chris Tarrant came to cover the story for the local news channel. And even though I'm not a man who could tell red wine from Red Bull in a blind tasting, I can see why. It just tastes – what's the word? Better.

As the flow rate suggested about a million litres a day were coming out of the ground, I figured there'd be enough for me, and that I could bottle what was left over and sell it in my farm shop. Yes, I know. Peckham Spring. But, as it turned out, much more complicated.

First, the water had to be captured in a tank before it had had a chance to see the light of day. Then it had to be fed down a new pipe to another tank containing a pump, which would shoot the water up yet another pipe to a plant room, where all sorts of witchcraft would be used to remove all the things that the report said weren't there in the first place.

From here the water would travel along a final pipe to

a wipe-down sterile room where it could be bottled. This was two shipping containers I'd welded together and kitted out with stainless-steel fixtures and fittings. I ordered the bottles and had labels printed saying 'Diddly Squat Water. It's got no shit in it'. Because I knew it didn't. And on the hottest day ever recorded in Britain, production began.

Now, unless you arrived here from Somalia via Libya and Sangatte, I'm guessing you've never been in a shipping container on a hot day. Don't try it. It was 52°C in there – so hot that I broke open every single bottle that rolled off the conveyor belt and downed the contents immediately.

Which is a bit scary, because before I could put the water on sale, I had to have it tested again, and, somehow, it's failed. I know it's clean at the source and I know all the pipework and filtration system is cleaner than the clean room at the Centers for Disease Control in Atlanta, Georgia. But a bacterium has managed to get in there, and now I have to flush the whole system out.

This will mean using detergent, which will eventually end up in the stream at the bottom of the garden. It'll then flow into the Evenlode and the Thames and then the sea, from where it will evaporate and be flung high into the sky before falling over Scandinavia as rain.

Which means that in a billion years from now, Lars

and Ingrid will drink a bottle of Norwegian mineral water, imagining it to be as pure as can be. But to keep us happy in the here and now, it'll actually be a litre of Fairy Liquid with some dead germs in it.

Help! I can't understand a word of farmers' agri-jargon

I didn't think farming would be especially difficult. I figured that man had been growing crops for 12,000 years and that after such a long period it would be in our DNA. That it would be relaxing. Monty Donnish even. I'd plant seeds, weather would happen and food would grow.

In my mind, then, farming would mostly involve leaning on a gate while munching pensively on a delicious Dagwood Bumstead sandwich, or enjoying a late-summer sundowner from behind the wheel of an air-conditioned tractor. It'd all be a festival of crusty bread, lemonade, fresh air and cider with Rosie. Followed by a cheery harvest festival and a big fat cheque from the EU.

I've learnt, however, that all of it is back-breaking and difficult, that there's never time for a ploughman's in the sunshine, that there's no cupholder in my tractor for sundowners or anything else and that to be a farmer you must be an agronomist, a meteorologist, a mechanic, a vet, an entrepreneur, a gambler, a workaholic, a politician, a marksman, a midwife, a tractor driver, a tree surgeon and an insomniac.

I am none of those things, which is why I spend every single evening with my nose buried in a copy of the countryside bible – *Farmers Weekly*. It's my new favourite thing.

I especially love the fertilizer and machinery adverts, because they all feature fifty-something men and they're all wearing checked shirts and zip-up gilets made from a material that exists only in agricultural supply shops. I want to buy everything they're advertising because it all looks so manly and proper.

The editorial is a bit different, though, because I can't really get my head round any of it. There will be a picture of some sheep, so I'll think, 'Ah. I have sheep. I must read this.' But after the second paragraph I have to give up and move on because I don't understand a single word.

I therefore switch to a piece about the new agriculture bill, but all I've taken in when I finish it is the sound of a voice inside my head saying, 'Concentrate, Jeremy. This is important.' The actual words? No. They've just swum about like fish.

I understand now how life is for people who think they might be interested in cars. They pick up a car magazine, and after five minutes they think that maybe the exciting front cover featuring a Porsche on full opposite lock was a con because the text inside seems to be about physics.

I can read about an electronic limited-slip differential and know what the writer means. I know terms such as lift-off oversteer and axle tramp and torque steer and scuttle shake and I even have a fairly good idea what the motoring writer Gavin Green meant in *Car* magazine when he said the then new Toyota MR2 suffered from 'tread shuffle'*. For most people, though, this kind of language is gobbledygook.

We see the same problems today with Formula One. The commentators don't translate tech-speak such as 'deg' for the viewers. They use it to demonstrate to the drivers and the engineers that they too are part of the inner circle. It annoys me – so, chaps, can you stop saying 'box'. And use the word 'pit' instead, because then people at home will know what the bloody hell you're on about.

This brings me on to the world of banking. Like a lot of people I have savings, and that means I occasionally have to speak with people called Rupert and Humphrietta. One said in a Zoom call recently that in the previous few months I hadn't 'shot the lights out'. I had no idea what she was on about. She then tried to sell me a 'product', which, it turns out, is only a product in the way that a casino chip on red is a product. I could be wrong, but I'm in no position to know.

* I actually don't know what 'tread shuffle' means.

I turn occasionally to the *Financial Times* for assistance on these matters, but, like the car magazines and the F1 commentary, it's far too complicated. Which is why I mostly end up reading the superyacht reviews in the disgusting but strangely engrossing 'How to Spend It' supplement.

I fear, however, that simplification isn't actually necessary in *Farmers Weekly*, because the readers don't need the jargon translated. When they read that ex-farm spot wheat values are averaging close to £176.50/t midweek, they know what the words mean and what the implications are. Me, though? Not a clue.

I have been writing these farming columns for six months and I have started buying all my clothes at StowAg, so quite often I'm stopped in the street by farmers wanting to know about the moisture content of my wheat or where I am on the idea of levying a carbon tax on farmers who finish their cattle after twenty-seven months.

I have therefore become very skilled at nodding and then suddenly remembering that I must get in the car and go away.

The worry is that I want to learn how to speak farming, but I have no idea how this is possible. I don't have a boss who can take me under his wing, and while I have a land agent, who's brilliant, he is even more un-understandable than *Farmers Weekly*.

I could sign up for a three-year course at what is now, hilariously, called the Royal Agricultural University in Cirencester, but by the time I'd finished learning how to drive a Golf GTI up the steps and how to get home from Cheltenham after a particularly pissed-up day at the Gold Cup, I'd be too old to lean on gates or climb the ladder into my tractor.

Muddling on isn't really an option either, because when our EU money dries up in January, it's very obvious farmers are going to have to adopt a much more scientific approach to survive with dwindling government grants.

I already don't know how a potato grows, but soon it won't matter unless I can use chemicals and boffinry to grow four billion of them. I shall therefore drown in tech I don't understand and can't afford.

I have turned to the internet, of course, and it is neatly split between two approaches. Fantastically simple nonsense written by and for failed City boys who have two acres and a lamb. And head-spinningly complicated equations written by people into chem-porn at Monsanto.

And in the middle of all this there's me, who wants to make good food, well. I think I'm not alone. I think there are a lot of farmers like me who are bewildered and even a bit frightened by what they must do to survive. And I think you, round your breakfast tables, should be worried too.

Because when you take the art and the history and the simplicity out of farming, I suspect you may end up with a lot of food that doesn't taste very nice.

November

Let British farmers take back control – of cheap, nasty food

The government's Eat Out to Help Out scheme was designed to prolong the life of Britain's seemingly doomed hospitality industry. But it had another effect. It allowed hard-up families to eat food in restaurants they could not normally afford.

I was at my quite expensive local when one such family sat down for lunch. And straight away they were unhappy because one of the items on offer was wild Scottish langoustine with burnt lime. None of them could understand why you'd want to eat something that had been burnt. Or why the owner would want to advertise his chef's incompetence on the menu. They were also 'disgusted' by the devilled kidneys, because who'd want to eat a kidney? And they had absolutely no idea what the hell gnocchi was.

When the food arrived they were even more cross because the steak was still bleeding, the chunky chips were nothing like the 'proper' chips they got from McDonald's and there were leaves on the plate. Actual bloody leaves. Eventually the father exploded. He did a lot of shouting, explained that he wouldn't be paying even his reduced share of the cost and drove away so vigorously that I half-expected the ladders to fall off the roof of his Vauxhall.

We have seen this kind of thing before. When Jamie Oliver started his healthy school-food campaign, mothers in Rotherham responded by turning up at the school gates with 'proper' food for their kids. Cheese slices. Crisps. Fizzy pop. And a nice bar of lard dipped in milk chocolate. Interestingly, one of the people who backed them was Boris Johnson.

An Irish court recently decided that the bread in a cooked Subway sandwich contains so much sugar that it cannot legally be described as bread. The *Sunday Times* food writer Marina O'Loughlin went off to try one and said the artificial taste 'lingers like herpes'. This may well be so, but the fact is that there's always a queue outside Subway. Many people like their bread to be sugary.

And all this causes me to arrive in a state of confusion at one of the many dilemmas vexing our beloved leaders. It's this. If we are going to do trade deals with America and Australia, we can't very well say, 'Oh, and by the way, if you want to sell pork here, can you give your pigs hot-water bottles and read them bedtime stories, because that's what we make the farmers do at home?'

When we leave the EU, the plan is that supermarkets will be able to import food that has not been produced to anything like the standards imposed in the UK. Put simply, you may well be eating chickens that have been doused in a bucket of chlorine to kill any of the bacteria they picked up during their short, cramped and miserable lives.

Horrific, you say. But hang on a minute. I know that if you are into fair trade, peace and veganism, or if you work on a submarine, chlorine is seen as a bad thing. But most normal people experience it only if they go swimming, and they like it, because if the water is teeming with chlorine it demonstrates that it's not also full of kiddie wee and chlamydia. They therefore won't mind if their chicken has been basted in the stuff. If it has then been infused with enough sugar and salt, it's possible they won't even notice.

The fact is this: food made to lower standards than we have in the UK will be cheaper. And, whether we like it or not, cheapness is what matters most of all to most people. Yes, everyone here wants to eat British food, but if an Israeli chicken costs 50p less than a chicken reared down the road, the Israeli chicken is going to go in the trolley and the British farmer is going to go on the dole.

There are noisy calls being made at the moment, mainly by the National Farmers' Union (NFU), for a trade standards commission to be set up. They want a panel comprising ecoists, animal enthusiasts and boffins to decide what can be imported and what cannot.

It is a noble ambition and I can see why it is supported by so many chefs, foodies and farmers. I support it myself. But what it will do is keep the price of food higher than it could be. So I have an alternative suggestion. Instead of ensuring British quality standards are imposed on food coming from abroad, could we not lower our standards to make farming here less expensive?

Most people think farmers pour industrial levels of chemicals on their wheat fields and shoot bees for sport, so why not simply do that? Because that way the cheapest chicken on the shelves would be British.

This would be good news for the farming industry, which would have fewer rules and better profit margins as a result. It would be good news, too, for the mothers of Rotherham, good news for the government, which could have its trade deals, and good news for that chap at the restaurant, who can spend the rest of his life feeding his fat kids with Bhopal-infused oven-ready British shit.

Yes, some people like good, well-made food. Every week in this magazine we see lots of recipes, and in the pictures there are always plenty of pine kernels and coriander seeds. There is definitely a market for this kind of stuff.

The butcher I use when I'm in London sells chickens at £28 a pop. I'm not making that up. And people buy them. People also come to my farm shop, where a jar of honey is just shy of a tenner. And many drool when I tell them that the wheat I grew was turned into flour at a mill three miles away and then into bread at a bakery at the end of the road. They like the localness of it all and are prepared to pay a premium for the loaves that result.

But let's not get deluded by this farm-to-fork guff. It's great and I'm going to do more of it, but I know that for every customer I have, Aldi has about 17 million. Because most people simply can't afford to eat what we call well.

I can see why you would want to make sure that all the food sold here has been produced with love, care and one eye on the environment, but that would be like me saying that we should ban crappy Hyundais and Peugeots in Britain because it would be much nicer if people drove Jaguars and Land Rovers instead.

The fact is that most families do not sit down around a table to eat supper. Many do not even have a table. They simply slam something from the freezer in a microwave and then wolf it down in front of the television. Or they call Deliveroo. Cooking? Only a quarter of us know how to make more than three things, and one of those things is bangers and mash. Which isn't really cooking at all.

So let's be realistic. If you say to someone who's filling his face while watching a soap opera that he should have paid a bit more to ensure the pig that made his sausages had a happier life, he'll stick his fork in your eye.

I wish the NFU well. I really do. But I fear they are selling an idea that appeals to about twelve people. Everyone else just wants some fish fingers.

WINTER

December

Why I won't be selling turkeys in my farm shop

The muddle-headed progressives in the left-wing media exploded with joy recently as they explained that farmers will soon be getting government subsidies only if they build down-filled igloos for the newts and knit snazzy jumpers for the trees.

They went on to say that farmers affected by this include Sir Dyson, Mrs Queen, the Duke of Westminster and Prince Khalid bin Abdullah Al Saud. And they're right. These people will be affected. But so will thousands of others who have just endured the worst farming year in living memory, thanks to the weather. And who now, thanks to Brexit and this subsidy business, face ruin.

This is what neither of the people who read lefty newspapers understands: that some farmers have Range Rovers and spend half the year spraying their subsidy cheques into Val d'Isère's cheese fondues, but the vast majority have to hold their trousers up with baler twine and burn their children at night to keep warm.

And what the lefties also can't understand, because they're too busy deciding whether to go to the women's lavatory or the men's, is that when England's farmers can no longer grow barley because in a climate-obsessed culture it just isn't financially viable, brewers will simply get what they need from Argentina, where there are fewer rules. Which means we haven't solved the environmental issues. We've just exported them.

Simple truths like that seem not to bother the bleeding hearts, though. They explained that farmers who didn't like the cuts in subsidies could sell their land to the poor, who of course are much better at everything than the rich.

Well, I've got bad news for you down there in Hackney and Islington. I shall not be selling my farm to a Palestinian refugee or anyone else for that matter. And, to make you even more angry, I shall remain in business by deploying the only thing I learnt at my very expensive public school: how to take a perfectly straight and simple rule and bend it so that it looks as if someone's spilt a bag of hairgrips into a bowl of Alphabetti spaghetti. 'That's not a nicotine stain on my fingers, sir. It's potassium permanganate.'

To limber up for this assault on the civil service and the left and George Useless at the bloody environment department, I'm going to try a new thing in my farm shop at Christmas, which is: not selling turkeys.

I do not keep turkeys, because they are even harder to feed than your wheat-, gluten- and dairy-intolerant teenage daughter who's just become a vegan. All they'll really eat are cherry trees and sunflower seeds and oats, but only if it's all dry and no other birds have stood on it. After you've kept your turkey warm and entertained and out of the wind for twenty-six weeks, you will have to kill it, and this is where the government steps in. Because you can't just hit it with a brick or shoot it in the face. You have to stun it first, by breaking its neck, unless it weighs more than 5kg, in which case you must electrocute it. And you are allowed to kill only seventy birds a day. No, I don't know why either.

It makes little difference to me, because although I have a licence to drive a car and another that allows me to operate a shotgun, I don't have one that lets me sell you one of my own turkeys in my own shop.

Not that you're going to want a turkey anyway this Christmas, because you'll be eating your lunch in a tiny group of three or four. And one's bound to be a vegan. And the other's going to be bird-intolerant. So it'd be silly to cook something the size of a blue whale.

What, then, is an alternative? What am I legally allowed to sell you that you might actually want to buy? A crow? A badger? A dragonfly? This is where you have to get creative. This is where you have to look at the rulebook and spot what's not there. And who better for inspiration than the French?

For centuries people all around the world have cooked bread and cows and fish, but the French decided that a small bunting called the ortolan would be more to their taste. So they tried it and then thought, 'Mmm, yes, but would it be better still if we caught it in a net and then put it in a box for two weeks, where the darkness will cause it to gorge on millet until it's dripping in fatty goodness?'

And, having decided to do this, they reckoned that they should kill it by drowning it in Armagnac, and then, after plucking it, they'd pop it under the grill for eight minutes and serve inside a buttered potato. Oh, and people would eat it while wearing a large napkin on their head.

In any normal country the people would rise up and say, 'That's stupid,' but in France everyone said, 'That's brilliant,' and I'm afraid they have a point. Ortolan is, by far, the nicest thing I've ever put in my mouth. When you bite into it the bones are soft like a sardine's. And the taste is like foie gras on a bed of – how best to describe it? Songbird, I guess.

Sadly, however, even though President Mitterrand loved the bird so much he insisted he had one for his last meal, by the late Nineties it had become so rare in France that serving it in restaurants was banned.

The end of the story? Nope. Because now, if you know where to look, restaurants will sell you a nicely buttered potato for €90. And you get, free, a bunting in it. 'But, monsieur l'inspecteur, we are not selling ze bird. We are giving it away. It clearly says so on ze menu.'

This, then, is what I'll be selling in my farm shop this Christmas: potatoes full of goldfinches and blue tits. It's bending the law, I know, but it'll be good practice for when the lefties and their new hero in No 10 try to turn the whole country into one big picnic site full of litter louts and wasps.

January

I'm not sure I'm cut out for farming

It's been an exciting morning on the farm. I've swapped 60 tons of hard core lying around in the yard for 90 tons of topsoil. Although when I say 'I've' swapped it, what I actually mean is that 'I've' been sitting at the kitchen table while a man in overalls has swapped it.

I was going to help. I got up early, full of vim and determination, and I pulled on my wellies and a thick coat and a fur hat with earmuffs, but it was one of those damp northeasterly mornings that can penetrate all known materials, including skin and bone. The thermometer read 1°C. But it felt colder than that. So I went back inside, took off my coat and my hat and my wellies and made some toast.

Ordinarily, in the depths of winter, arable farmers are to be found turning their subsidies into glühwein and hot cheese in the Alps, but as that's not possible this year because of the Covid, they are all stuck at home, doing jobs that have needed to be done for years.

And for the first time since I began eighteen months ago, I'm wondering whether my heart is really in the whole farming malarkey. While I very much enjoy driving round the farm in my Range Rover looking at stuff, and doing a spot of light cultivating on a glorious autumn evening, I am not even remotely inclined to go out in the middle of January to mend a gate.

Furthermore, I am not able to. You know those little sachets of wasabi you get with takeaway sushi? Everyone else can get into them with no problem at all, but I cannot. Nor can I follow even the simplest instruction booklets. And then there's my phone. When it updates itself in the night, I have to throw it away because it's different. And different means worse.

All this means that when I'm presented with a gate that's not attached to a gatepost any more, I'm stumped. Usually I'm so stumped that I'll spend fifteen minutes wondering loudly why the hinge fell off, and then I'll go back inside for some toast.

Sometimes, though, if the broken gate is sheltered from the wind and it's not 1°C, I'll look for the detached hinge, and then, when I find it five or six feet away, I'll wonder loudly how the bloody hell it got there. Then, because I'm no longer able to bend down and pick it up, I'll go inside for some more toast.

Most people can bend over to pick stuff up without thinking, but it's no longer possible for me. If I bend at the hips, I get a jarring pain in my kidneys, and if I bend at the knee, I know I will not be able to get up again. This is a problem, because the ability to bend over in farming is as important as the ability to do strangling in the special forces.

The knee issue also means I am no longer able to jump off things. And that's the second thing farmers must be able to do. Jump off trailers and walls and gates. Bending over and jumping off things – I'd say that's 80 per cent of a farmer's life. The other 20 per cent is going to the hospital to have your arm sewn back on.

Happily, my arm is rarely in danger of coming off, because I still have no clue how to attach any machinery to the back, or front, of my tractor. So, instead of trying, I choose to sit at the kitchen table reading the papers, while munching gently on a well-buttered pikelet.

There is, however, no getting round one job. No matter what the weather's doing, I have to fire up my six-wheel-drive ex-army Supacat, attach the ex-army trailer using an extremely manly Nato hitch and head into the woods for firewood.

Firewood used to be a simple thing, but now the government has decided to complicate matters by banning the sale of wet logs so people don't burn them. Quite why anyone would want to try to burn a wet log, I have no idea. It'd be like trying to stay warm by burning a wet towel or a wet dog.

But, anyway, as I understand it, I can no longer use soggy, mossy logs that have been lying on the ground, and instead – for the sake of the environment – it seems I have to chop down trees. Naturally I'm not very good at it.

In my head a chainsaw is a tool of the gods. No one picks a fight with someone who's revving a Stihl. Brandish one and you're the most powerful person in the room, unless someone has an AK47 – and even then it's by no means a foregone conclusion.

And yet, when I have one in my hands, I always have the sense that I'm the one most likely to be injured. I am in constant fear, for example, that the chain will come off and cut me in half. Or I will slip, and then I'll be in A&E with all the other farmers, having a limb sewn back on. Chainsaws terrify me even more than sharks and quad bikes.

And if I take a brave pill and get cracking, I will only ever get halfway into the tree trunk before it jams, and I'm not able to unjam it, because all the safety equipment I'm wearing means I can't see anything. At a rough guess I'd say 20 per cent of the trees in my woods have chainsaws stuck in them.

Sometimes, though, I will get a tree to fall over, and then, after I've climbed out of the branches and repaired my lacerated face, I have the job of loading it into my trailer and trying to get out of the wood. I can't do that either.

Thanks to their ability to lock all the wheels on either side, just as a tank can lock its tracks, Supacats can turn in their own length, which makes them incredibly manoeuvrable. But when a trailer is attached, the turning circle is measurable in light years.

This means I have to cut down more trees to create a path back to the world, and because there's no more space in the trailer, they have to lie on the ground becoming wet and illegal.

And do you know how long a tree lasts in my firepit? Well, if it's a good size, I'd say: 'Less than an hour.' And then it's back to the wood for more deforestation and devastation. If only we could still use coal.

But we can't. And when the day comes when we aren't allowed to use oil and gas either, the only way we will be able to stay warm is to go for a brisk walk.

I did that the other day, and in a field I thought I'd planted with grass I found thousands and thousands of radishes. Which on reflection may be adolescent turnips. Like I said, I may not be cut out for farming, because either I don't know what I'm doing or I can't be bothered to do it.

I don't need Joe Wicks.
Farming is keeping me fit

Dry January went well. Sales of wine increased by a third and of beer by nearly a half. And that's just what people drank while they were making cocktails. Tequila sales rocketed by 56 per cent and rum was up by a whopping 64 per cent.

I understand this philosophy and employed it myself in the first lockdown. Every evening, as the sun sank over the beech trees into yet another soundless crimson goodbye, I'd open a chilled bottle of rosé and sit listening to the wood pigeons until I decided that what I needed most of all was a mojito. So I'd sway around the kitchen garden, collecting mint, and then, to soften the blow on my stomach lining and liver, I'd nibble on fresh watercress from the stream until it was time for a swift Baileys and bed. God, they were wonderful days. Quiet days. Happy days.

However, they did take their toll. When we were allowed back into the world, I was so fat I looked like Ayers Rock on a unicycle.

I couldn't bend over to do up my shoelaces, I walked as though I'd had a trouser accident and my knees ached constantly from the sheer effort of keeping my landmark-sized torso upright. Which is why, when this lockdown started, I adopted a different strategy.

This time I decided I'd emerge at the other end a new man. A better man. People would stop me in the street assuming I was Iggy Pop or Willem Dafoe. I'd be like those fell-farmer chaps you see on *Countryfile* who are ninety-five years old but can still run up a Scottish mountain while carrying a sheep. In short, I would replace booze with exercise.

If you go to a gym, you pick things up and you put them down and you look at yourself in the mirror and then you go home. Whereas if you go and do proper old-fashioned farming, with proper old-fashioned tools, you come home at the end of the day having achieved something.

And don't say, 'But I haven't got a farm,' because, let's face it, you haven't got a gym either. You pay to use someone else's, and if you pay me I'll let you come to Diddly Squat and help me chop logs. Actually, this may be a neat solution to the financial problems caused by dwindling agricultural subsidies. Farmers can rent axes to attractive young avocado enthusiasts and send them off into the woods.

Now is the best time of year, because the seeds are in the ground and it's too wet and windy to do any spraying. Farmers, therefore, are forced to turn their attention to muscle-building maintenance, mending gates, replacing rotten fence posts and repairing walls that the badgers have knocked down. So if your name is Arabella or Camilla and you really want some taut abs, send me a cheque for a hundred quid and I'll set you to work.

To make sure the idea worked, I decided to do hedge trimming. Normally I use an enormous and ugly machine fitted to the back of my tractor, which goes through a hedge and everything in it like a power drill through a bag of muesli. That's why we have to trim hedges now, before the birds start nesting.

This time, however, I'd be doing it manually. I therefore needed a tool of some sort, and that was good news, because it meant a run to StowAg, which is the best shop in the universe. If you want something ugly and practical and farmerish, this is where you go. If this place wore a shirt, it'd be Viyella, and if it had shoes, they'd be as stout as they were brown.

I was distracted at first by the pig troughs and horse buckets and meaty-looking chainsaws, but eventually I found myself among the branch-cutters. There were many to choose from, but I'm a man who equates weight with quality, so I went for the heaviest.

Back at home I pulled on my gym kit: a tweed coat with twenty 12-bore cartridges in each pocket, and a pair of wellies. And off I went into the big green, to trim a hedge that in the past year had enveloped a little-used gate.

Here's how it works. You find a branch that has grown over the gate, follow it back into the hedge, insert the cutting tool and squeeze the handles. Then you grab the severed branch and, after the thorns have torn chunks out of your hands, you walk back to the farm, get in the car and go back to StowAg to buy some sturdy work gloves.

Soon I was hard at it. Bending, stretching, squatting and squeezing with all my strength to go through the bigger branches. My arms ached from the effort of lifting my overly heavy tool, my glutes were throbbing and my heart was beating nineteen to the dozen. It was minus 1°C out there, but my face was red and in my tweed coat I had moob sweat. I also had a pile of branches and, most importantly, a fully functioning gate. Can Joe Wicks say that after one of his workouts? Can Mr Motivator?

That afternoon I decided to knock in some fence posts, and that's even better. Again, there's a machine that can do it very simply; you just sit in the warmth of the tractor and push a button. But, again, I elected to go old skool and used a fence-post knocker. It's like a section of steel drainpipe, sealed at one end, with handles attached on either side.

Operating it is easy. You position the pipe over the post and, summoning all your strength, use the sealed end as a giant hammer. I've seen some fairly brutal-looking workout equipment in gyms, but nothing gets close to this. Using a Force USA Monster G6 power tower is like angling on the Shropshire Union Canal. Building a fence is deep-sea fishing for marlin. It's why you will never see a fat fencing contractor.

I did two posts and my arms had had it. They hung by my sides as if they had been filled with zombie spice. And I still had the long trudge up what feels like the steepest hill in England back to my farm, with all that lead in my pockets and with mud-caked wellies that weighed 200lbs each.

That night I was feeling so righteous and so full of fresh air and so healthy that I didn't want any wine or beer. I didn't even want a mojito. Instead I drank water from my spring and, using bread that had been made from my own wheat, made a tomato and ham sandwich.

I've always seen my farm as many things: a place of great beauty, a fun business and, if I'm honest, a good way of passing on wealth to my children without the taxman getting involved. I never really saw it as a wellness spa, though. But that's what it has become. And I recommend it because, like I said at the start, dry January went well. I enjoyed it.

February

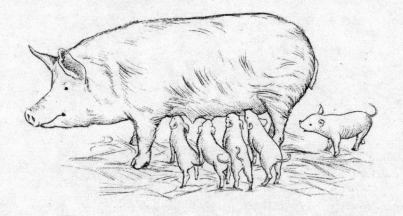

Will my pigs be saved from the frying pan?

My signature dish is a pork and pepper pasta. Though when I say 'signature dish', what I mean is 'the only thing I can cook'. Here's the recipe. Coat some diced pork tenderloin in flour, fry it in my own vegetable oil with some chillies, green peppers, onions and mushrooms, and then, after adding stock, a splash of cream and seasoning, serve it on a bed of fusilli pasta. It's nice.

And it's not the only pig-based food I enjoy. I'm very partial to crackling and sausages, and I love ham with broad beans in a parsley sauce nearly as much as I enjoy seeing a suckling pig spinning slowly over an open fire with an apple in its mouth.

In a restaurant, I'll happily order pig's cheeks, unless it has trotters on the menu, in which case, I'll have those instead. And as we know, a bacon sandwich made with sliced white bread and Heinz tomato ketchup can cure hangovers, vegetarianism and even the common cold. Never in the field has so much been given to so many by one animal. And as a result, I'm thinking of becoming a pig farmer.

There are other reasons too. Pig foraging disturbs and invigorates the soil, causing roots, bulbs and seeds to germinate. And their manure is teeming with goodness, which means I'd spend less time driving about in my eight-litre tractor, showering the farm with chemicals. So, they are not only delicious and versatile, but also good for the environment. Perhaps that's why so many eco-people look like Arnold out of *Green Acres*.

Pigs can also be used to settle neighbourhood disputes. Let's just say you're a farmer and someone whose house adjoins your land is annoying in some way. You could ask the courts for help, or the parish council, or the local newspaper, but in my experience these things never work. It's much better to put pigs in the field next to his gaff and say you won't move them unless he backs down.

I really couldn't see a downside to my pig plans, but before diving in, I decided to dip a toe in the water and start off with a couple of Kunekunes. This breed was on the verge of extinction in the Seventies but since then it's become the trendy, must-have labradoodle of farmyard animals, and everyone with half an acre now has a couple truffling around in the brambles.

It makes sense, as they come with built-in fur coats so they can live outside all year round and as they are the only true grazing pigs, they can survive quite happily on nothing but grass and vegetable peelings. Also, they are surprisingly cute.

Maybe this is because pigs are actually very similar to human beings. Their organs are laid out just like ours, which is why they are often used for medical research, and their flesh is so similar that weapons specialists often use pigs to test the effectiveness of bullets. What's more, pigs know to disregard their own faeces, they have long eyelashes just like Twiggy, many can speak forty languages and a recent study found they can operate a computer joystick. Pigs can even recognize themselves in a mirror.

They are also, I've learnt, extremely good at escaping. I put them in a field that was used last year to grow vegetables, so it's teeming with discarded chard and potatoes and beans. It's pig heaven. I even bought them a nice house with a window and angled it so they have a lovely view down the Windrush valley.

But they obviously hate it in there because on the very first day, both of them charged the electric fence and were gone. And have you ever tried to herd pigs? It's like trying to sweep air. And if by some miracle you do get them back in the right general area, they take one look at the orange string that gave them an electric shock on the way out and that's it – they're gone again.

To make matters worse, pigs can travel, I've discovered, at several hundred miles an hour. It took four hours and six people to get them back in their pen and thirty minutes later, they were out again. It was cold and dark and sleeting, and this time one of them went into a hedge and refused point blank to come out. The other kept biting my leg.

Yesterday morning I noticed they'd turned their house into a vaulting horse and in the afternoon I received a call to say that one of them was riding a motorcycle down the fence line between Germany and Switzerland.

I've now built a proper wooden fence and when I took them some vegetable peelings, one of them was sitting in the corner of his house, endlessly throwing a baseball against the far wall, while the other was making what looked like a glider.

All farm livestock will try to escape. But usually their attempts are opportunistic and badly thought through. My sheep will saw themselves in half to get through a gap in the hedge, my hens will risk an encounter with Mr Fox as they make a break for freedom and my trout will wriggle across ten feet of grass to get from their perfectly lovely pond into a nearby bog.

The pigs are different. They understand searchlight patterns and always go in different directions when they're out. I haven't given them names, for obvious reasons, but if I did, I'd call them Stephen and Stephen. After Fry and Hawking. They're that clever.

And yet, in some ways, they are like children. When I feed them, the big one always stands in the trough so the smaller one can have nothing. And God, they fight. Constantly. Usually over whose turn it is to play with the telescope they've made.

This worries me because what I have in the field, when all is said and done, is 420 sausages. Pigs are a business. You get a sow pregnant, she has a dozen piglets and you can either sell them at eight weeks for about £50 each. Or you can keep them to adulthood and sell them for £600 a pop. That's not profit, though.

You've got to factor in the cost of feeding them, and housing them in what's basically Parkhurst and the cost of killing and butchering them, but you should clear £200 a pig. It's not dog-breeding money but it's better than a kick in the face.

The trouble is that I like my Kunekunes. I like the noises they make and their spirit and even their scrunched-up faces. I had no real problems taking my sheep to the abattoir and even less hawking a trout from the pond when I'm hungry, but I don't think I could eat the pigs. They've put me right off pork, in fact.

Tomorrow then, I'm going to chop up some swede, celery, mushrooms and onions and pop them in a slow cooker with some browned cow. And I shall serve it six hours later with some buttered mashed potatoes. It'll be my new signature dish.

And then, after I've fed the pigs and tickled them behind their ears, I'm going to see if it's possible to make bacon out of hens.

Birdwatching is now the highlight of my days

The Royal Society for the Protection of Birds was formed towards the end of the nineteenth century in a bid to stop rich women decorating their fur coats with the feathers from great crested grebes.

I find this surprising because, back then, there were many problems in the world. The Greeks and Turks were at war with one another. American troops were slaughtering people who at the time were called Red Indians. Britain had just suffered its first ever terrorist attack (at the Greenwich Observatory), the Swedes had discovered a link between carbon dioxide and global warming, and all over the world people were dying in vast numbers from the bubonic plague, leprosy, smallpox and cholera.

But despite all this, some idealists in Manchester said: 'Yes. But what matters most of all is the way toffs are using grebe feathers to decorate their coats.'

The RSPB has been at the vegan end of the political spectrum ever since and must now be viewed as the Labour Party's Luftwaffe. I'm surprised it hasn't adopted the Palestine sunbird as its emblem.

However, despite this, they do organize the rather excellent Big Garden Birdwatch every year. It's very simple. People are asked to sit at the kitchen window for an hour, making a note of all the birds that drop by. It's a good way of seeing what breeds are up and which are down, and it's become the biggest bird survey . . . in the world.

It's so big that before it began this year, an extra 300,000 fat balls were bought online. Meanwhile, the sale of seed skyrocketed and in my local town it was impossible to find any peanuts on the shelves at all. I had to buy pistachio nuts instead.

Yup, I took part. I like birds; always have. I've littered my farm with kestrel and owl boxes, I've planted the margins with a turtle dove mix to try and attract one of the most endangered birds we have, and in the cold snap I spent hours trudging round one field as I'd had reports that a lapwing had moved in.

On top of all that, I don't trim the hedges, so they now look like Germaine Greer's lady part, and I've put so many insect-friendly flower strips through the big fields that from the air it looks like they're made from corduroy.

And I like to think it's working. Back in the summer, I saw a flock of goldfinches, and in one uncut hedge there are more than a hundred yellowhammers. This is a bird the RSPB once said was in grave danger, thanks to the motor car. Really? Are they suggesting the yellowhammer is less able than other birds to get out of a car's way? Or that 'motorists' have made it part of some weird game to keep the kids amused on long journeys. 'Hey kids. Look what just smashed into the windscreen. That's ten points!!!!'

The Big Birdwatch, however, is not interested in what's on the farm, only what's in the garden, so having peeled the pistachio nuts and loaded up the bird table with grease and lard and grain from last year's harvest, I sat back with my monocular and my notebook and some reference books to start the count.

First up was a robin followed shortly afterwards by a blackbird that the robin chased away. It then proceeded to chase away two sparrows and a pied wagtail. It was so busy chasing everything away, in fact, that so far as I could tell, it never actually ate any of the food I'd prepared.

There was a similar problem on the hanging basket of nuts. A blue tit arrived and was quickly joined by a great tit. There was space for both of them to eat in peace and harmony, but no. They decided to have a fight. On and on it went, in a blur of flapping and pecking and squeaking.

In the olden days when Sir Attenborough told us interesting things and wasn't just a stuck climate change record, he would say that all creatures are careful not to burn more calories than could be provided by available food. But those two tits would have needed a McHappy Meal with fries to put back what they lost in that scrap.

In my mind, I'd see a nuthatch and a nightjar and a green woodpecker but after half an hour all I had was the angry robin, which had secured the perimeter and was now doing sentry patrols, and the two fighty tits.

We see similar character traits in birdwatchers as well. In New York, a full-on civil war has broken out between twitchers, with one group saying rare bird locations should not be publicized on social media and the other happily tweeting the GPS location of a snowy owl they've just seen. Everyone's so busy hurling insults that no one would notice if a dodo appeared in Central Park.

Meanwhile, back in Chipping Norton, I'd given up with the garden itself as the tits were now re-enacting a scene they'd seen in *Avengers: Endgame* and the robin was setting up a minefield. Instead, I'd turned my head to the sky but here too there was an issue, because although I know a bit about ornithology, I have to admit that one small brown bird looks pretty much like all the other small brown birds when the backdrop is bright and it's doing 40 mph.

Sir Starmer's air force has produced a handy online guide to help us identify a bird we've seen. But the only information I had was 'it was brown and in the sky' so I was rather stumped.

I began to speculate how much more fun a garden watch of this type would be if I lived in the Seychelles, where there are terns that appear to be made from a translucent porcelain. Or Papua New Guinea, where there are crows that look like luminous satellite dishes, and the *Disraeli Gears* album cover and Ed Sheeran.

Our crows are much more boring to behold but that said, they are extremely clever. It's known that pigeons can tell whether you have a gun or not but crows go one step further. They can tell what sort of gun it is. 'Ha. That's only a .410. He'll never be able to hit me with that.'

I proved this forty minutes into my Big Garden Birdwatch by going outside with a 20 bore. The pigeons scarpered immediately, but the crows just laughed at me, knowing they were well out of range. They were correct too, and having proved it, there were no birds in my garden at all.

Here's the funny thing though. After the survey weekend finished, I continued to put seed and nuts and fat in the garden every morning, and I now spend well over an hour every day staring through my monocular at the comings and goings. This morning a wren came. It scampered down the wall on its funny little legs but before it reached the little pile of seed, the robin arrived out of the sun and scared it away.

It is a phenomenally violent little thing, fiercely protective of its patch and not afraid to take on birds that are twice its size. Small wonder it was recently voted Britain's favourite bird.

SPRING

March

Can I stop my farm's floodwater from soaking the neighbours?

Storm Darcy, or Brian, or Enid, or whatever the last bit of bad weather was called, is only part of the perfect storm that's currently causing half of Britain's farmers to think about selling up and doing something more rewarding. Like being a town crier. Or a lamplighter.

They are no longer allowed to spray what's necessary on to their crops, which means their barley is 9in tall and the same colour as a Royal Navy destroyer. Then there's Brexit, which has screwed half their markets, and Covid-19, which means their fields are full of bewildered townies shouting 'Fenton!' at the top of their voices as their dogs tear around eating sheep and knocking over walls.

The government is not helping either as farmers have been told that if they want any income at all in future, they're going to have to stop growing food and turn their land into a sort of eco theme park for the uninterested children of Britain's delusional *Guardian* readers.

On top of all this, vegetarianism has gone from being a niche-interest activity for sixth form socialists to a mad nationwide craze – like clackers and pet rocks. Farmers, then, spend all their lives making their cows happy and are then lambasted from all quarters for having cows in the first place. Small wonder that on average one farmer commits suicide every three weeks in England and Wales.

And I suspect that number is about to become even more troublesome because we were all promised global warming, which doesn't sound so bad. Everyone likes a warm day. But what we got instead is a nationwide soaking. These days, barely a month goes by in the winter without some rainfall record being broken. February 2020 was the wettest ever. And in parts of England and Scotland, this January was twice as wet as normal. That's irritating for most people but for farmers it's catastrophic.

And to make matters worse, our esteemed leaders have decided to design floodwater defences so that farms are sacrificed to protect the three-piece suites of people in towns and cities. And are farmers compensated for this? Ha. You're having a laugh. There are proposals but for now they simply take one for the team and then the team throws excrement at them for selling meat.

I've been told that to help I should dam the streams on my hilltop farm because water held in ponds here is not able to enter the houses of people who live downstream.

And being a good citizen, I've spent the past year doing just that.

I like damming streams. As a kid, we'd holiday in Swaledale and I'd spend all day in the river outside Muker, trying to block the water's path with stones and rocks. It was an entirely pointless pursuit then and it's an entirely pointless pursuit now.

Water is relentless. It's a Terminator. It absolutely will not stop looking for a weakness, and when it finds one it's not happy to escape in slow motion. It wants to get cracking in one big rush. It took me six attempts to block the path of one stream and I succeeded in the end only after buying ten tons of stone, fourteen big sacks of cement, two men, a sluice gate, a digger and a massive pump.

For the next stream I got super-serious. I pulled on my Hoover hat and went berserk, creating a scene that Sam Mendes could have used if he'd decided to make *1918*. Huge escarpments were created, mighty trees were felled, the air was thick with diesel smoke and the sound of hydraulic power waging war with nature. The lake that all this created is 70ft long and maybe 30ft wide, and I felt proud because the water it contained could not be coming out of the plug sockets in your house.

Sadly, however, I was underthinking the problem because Oxfordshire is currently a building site. When I first moved here twenty-five years ago, the half-hour drive to the motorway was pretty and green and full of leaves. Now it's like driving through Surrey. Every village is ringed with new-builds and the city of Oxford is now bigger than Los Angeles.

So let's do some maths. In January, 2.2in of rain fell in Oxfordshire; so, if the roof of your house is 20ft wide by 50ft long, this means that 1,142 gallons of water fell on it. That's a lot.

And there are plans to build 28,000 new houses in and around the city in the coming years. Which means that every year, more than 300 million gallons of water that would normally seep into the earth gradually are cascading down gutters into drains and into rivers. Which means they'll become raging, vindictive monsters.

And remember, in addition to the problem caused by roofs, you've got the driveways and the roads and the decked gardens to think about. Britain's getting wetter and soon there'll be nowhere for the extra water to go.

I see the effect already on my farm. I recently built a small barn. It's maybe 40ft long by 80ft across. And outside it is a newly concreted yard. It all looks very smart, but in January more than 3,000 gallons of water that should have seeped through the brashy soil shot through the drainage system and straight into my streams.

I did a flow test the other day and couldn't quite

believe the findings. In the summer about two million litres of water were flowing down one stream each day. Last week it was handling five times that amount.

Ordinarily there are about fifteen little springs on the farm. Now there's one big one. Water is leaking from literally every pore. And the effect on my new big pond has been dramatic because the 4-in outlet pipe simply can't cope. Water levels consequently rose until the banks were breached, and that meant my trout escaped. So if you're reading this in Oxford and one of them swims into your living room next week, can I have it back?

In the meantime, I've had an idea that may help farmers and landowners in these desperate times. Britain has always been useless at managing its water. We live in one of the wettest countries on Earth, but somehow every time there's a two-day dry spell we are told to shower with a friend and not use hosepipes. This may have something to do with the fact that during the Sixties and Seventies, we built all our reservoirs in the north because we assumed people would move there for work. Only to find that everyone moved south, where there are hardly any reservoirs at all.

So let's build some. The government doesn't want us to grow crops and the nation's vegetarians want cows to roam free like their cousins in the Serengeti, so let's dam our valleys and grow water instead. We can rent it to idiotic wild swimming enthusiasts in the winter and then sell it in the summer to gardeners and people who are dirty. Everyone wins.

Why my farm shop is causing controversy in the Cotswolds

It seems that whenever we turn on the television these days, we are treated to the uplifting spectacle of a pink-cheeked countryside person birthing lambs and nurturing rhubarb. We have *This Farming Life*, *The Farmers' Country Showdown*, *The Great British Countryside*, *Countryfile*, *Escape to the Country*, and soon, because I've spotted a gap in the market, *Clarkson's Farm*.

And all of this is before you get to Channel 5, which has combined its love of Yorkshire and the royal family to give us *Our Yorkshire Farm*, *This Week on the Farm*, *City Life to Country Life*, *Build a New Life in the Country*, *Ben Fogle: Make a New Life in the Country*, *All Creatures Great and Small*, *The Queen's Farms in Yorkshire*, *All Creatures Great and Small with Ben Fogle*, *Ben Fogle on the Queen in Yorkshire*, *Farming in Yorkshire*, *Farming in Yorkshire with Princess Anne and Ben Fogle*, *Escape to the Farm with Kate Humble* and *The Duke of York's Yorkshire Puddings Made with Sarah's Ferguson.*

All these shows are designed to do the same thing: to show Tube-and-bus commuter people in towns and cities that life in the rural uplands of Britain is rosier and more sunlit.

Hmmm. When you watched Simon Pegg and Nick Frost's *Hot Fuzz* movie, you will have assumed that it was fiction, because obviously a bunch of respectable sixty-something shire people would not go around murdering those whose houses and clothing didn't match traditional village values. But I'm not so sure.

Quite recently we heard news from East Sussex about a rewilding enthusiast who's been told by his local council to tidy up the 'eco-paradise' he has created in his back garden. He said he's 'absolutely devastated' by the decision and we can only hope and pray that the poor chap doesn't turn up soon with a bit of cathedral in his head.

The problem is simple: in a village, most people are charming and happy to smile and wave at the appropriate time, but there is always a tiny minority who are bitter and whose mouths look like cats' anuses. These people are usually called 'parish councillors' and seniority in this weird world is achieved by having lived in the area for a long time. That's it. So, if you are the sort of person whose horizon is located on your nose end, and you've never been further afield than Chuntsworthy, you're the village elder. You're Hiawatha.

I covered parish council meetings for many years as a local newspaper reporter, which is why I always thought

The Vicar of Dibley was a documentary. Because they really were that small-minded and mad.

In the Yorkshire village of Brinsworth, councillors once spent forty-five excruciating minutes deciding that they needed a new water jug for their meetings, and then another forty-five minutes arguing about whether it should be made of glass or plastic.

Parish councils are clubs for people who want everything to remain as it was in 1858. If you move to the village and complain about a local tradition, say, or hanging witches from the maypole, the parish council will completely ignore you. If you decide to make your garden as messy as nature intended, you will be speared through the neck with a pair of shears. And if you open a farm shop, you'd better get ready for the Four Horsemen of the Apocalypse because, trust me, those babies are coming for you with burning knives on the wheels of their chariots.

My farm shop is tiny but it seems to have landed in this part of the Cotswolds like a nuclear weapon full of sarin gas. Sometimes I wish I'd built a mosque instead. Or a bypass. It would have been less controversial.

We all know that planning regulations are necessary and we all know that parish council enthusiasts are entitled to register their opposition, but there are some people in the countryside who literally do nothing all day long but object to stuff. They are made entirely from a blend of skin and hate.

If you've ever tried to build a spare room above a garage or chop down a tree, you'll know what I mean.

My shop had only been open a few days when we received a stern letter warning us that our rather lovely ice cream had been made from the juice of cows that lived eight miles away, in Gloucestershire, and that this contravened a clause that said that we could only sell produce from West Oxfordshire.

Since then we've been told that the roof is the wrong colour, that the sign is 0.3 of a metre too wide, that we aren't allowed to sell teas and coffees, that the gingham covering on the straw bales contravenes Covid regulations, that the car park is a road safety hazard, that the sausage rolls are wrong in some unfathomable way, and that if we were allowed to sell beer, yobbos would come and urinate in the graveyard.

All of which just goes to show how out of touch these guardians of the nineteenth century are. Because, these days, if you really want to attack something (or a royal family) you have to accuse it of causing mental health issues, or say that it's racist. It's not enough simply to say that the milk is from the wrong postcode.

It's strange. I lived in London for many years and apart from one time when I may have been using a bin bag full of rubbish as a football in the middle of Fulham Road at two in the morning, I don't recall a time when I ever fell out with a neighbour. I think that because people in a city are forced to live cheek by jowl with one another, they go out of their way to be stoic and tolerant.

In the countryside, though, contrary to what you see on TV, it's a very different story. If there has never been a farm shop, then there should never be a farm shop. Especially if it's run by someone who, like me, has lived in the area for only twenty-five years. I bet when Alexander Fleming invented penicillin, the village elders ran around saying that diarrhoea had been a part of rural life for hundreds of years and that they wanted to make sure it stayed that way.

The annoying thing is that only a tiny number of people object, but you never really know who they are, so you end up distrusting everyone and then sneering at them in the paper shop. And I wonder. Could that not be solved by posting the names and photographs of those who've objected on the community notice board? Or maybe we should make them walk round in Day-Glo baseball hats. I think that would help make village life as pleasant as you think it is.

April

Brexit red tape is sowing the seeds of disaster

As we know, the nation's Brexiteers are now running around with a smug look on their faces and Union Jacks on their mobility scooters, telling anyone who'll listen that if they hadn't won the referendum, we'd all be painting red crosses on our front doors and throwing Granny on the cart because she'd just coughed.

'Look,' they scream, priapic with delight, 'even the Germans have not been able to organize themselves thanks to the bureaucratic monster that is the EU, whereas here, where we are free and agile, every man, woman and child has now been vaccinated with a proper British jab.'

Even I will admit that the vaccination programme has been a remarkable success, chiefly, I suspect, because we bypassed the monstrously inefficient Department of Health and the almost completely useless NHS – the organization, not the doctors and nurses – and gave the job of arranging everything to a small team of people who were properly motivated by what Boris called 'greed'. Because that's what it was.

However, already I have come face to face with a major downside of leaving the EU and on balance, I'd rather have Covid . . .

Last year, global warming stopped being an issue that affects only unpronounceable islands on the other side of the world, and arrived in all its oppressive majesty in the UK. God, it was hot. Hot and wet. Which, to paraphrase Adrian Cronauer, is OK if you're with a woman, but not so good if you're trying to grow wheat.

From my kitchen window, we seem to be having the same problem this year. I'm writing this at the end of March, and I'm on the highest, windiest and coldest farm in West Oxfordshire, and already the oil seed rape is starting to flower. There's not as much as I would like – thanks to a surge in the number of pigeons – but it's yellow already. And that's weird.

So because the weather is obviously changing and because there's literally nothing we can do to stop it, I decided to adapt.

This is something coral should think about doing. Instead of sitting off the coast of Australia, bleaching and moaning about how the water's too warm, why doesn't it move to the Humber estuary and grow there? It's the same story with all those elephants we see in nature programmes, mooching about in muddy puddles, wondering where the river's gone. Come on, guys. You're supposed to be intelligent. So if water is your issue, move to Manchester.

Farmers in Britain must do likewise. So, as I now have many weeks of agricultural experience under my belt, I decided to adapt to climate change by growing durum wheat instead of the normal milling variety that's used to make bread.

Durum is a 'hard' wheat that was developed by man around 10,000 years ago and is perfectly happy growing in harsh, dry and hot conditions. So if that's what we are going to have in the UK from now on, that's what we will have to grow. And I was going to be one of the first out of the blocks – I'd be the kid in Formula One who changes his tyres before anyone else. I'd be ahead. In the lead.

Globally, only around 6 per cent of the wheat grown is durum because it's not what you'd call user-friendly. It's hard to mill and the casing is brittle so you get a lot of bran in the mix. Plus – and this is farmer speak – it loses its Hagberg very easily, so unless you get it out of the ground and sold very quickly, you end up with a shed full of pheasant food.

The man at my local flour mill was delighted, though, that I was going to give it a bash because in the UK in recent years there's been a surge in demand. And it's easy to see why. Not only is durum flour used by the middle classes to make pasta but in addition it's needed to make flatbread and Levantine dishes such as tabbouleh, kashk, kibbeh and the bulgur for pilafs. Which is pretty much the staple menu in every Huddersfield takeaway joint these days.

As a result of all this, I was feeling very smug. I had a new crop that could cope with hot, dry weather, and it would make flour that's jolly popular with those who enjoy a doner kebab after a pint. That's a double top.

As you can't easily buy durum seed in Britain, I placed my order, through a complicated chain of middlemen, with a French seed breeder in the Rhône Valley. And very soon, three tons of the stuff arrived in Calais, where it got stuck in a jungle of red tape.

The French customs said it would not be released until they were given the consignment's EORI number, and no one on this side of the Channel had the first clue what that was. And there was no point asking the French for clarification because all you get is the Gallic shrug, a universally recognized symbol of complete uninterest. Tinged with a hint of 'Well, you shouldn't have left the EU, should you.'

I had just spent £45,000 on a snazzy new Weaving Sabre Tine seed drill and after a lot of swearing and broken fingernails and calls to my farm manager to come and help, it was attached to the back of my fuelled and serviced Lamborghini tractor. I was ready to get out there. But I didn't have the seed.

The weeks passed and the weather got hotter and hotter. And as the thermometer climbed past 24°C, I started to worry that I'd missed the boat completely, and wouldn't be able to plant it even if it did turn up. I was so cross that I drove over to see a Brexiteer neighbour yesterday morning and called him a c***. I did. I pulled up, called him a c*** and then drove home again.

And I'm not alone. You try buying flower seeds from Holland these days. Or exporting corn and straw. I know we keep being told that traffic between the EU and Britain is barely affected by Brexit but from where I'm sitting, that sounds like nonsense.

Happily, yesterday afternoon, my bacon was saved by Simon Bates. I'm fairly sure it's not the Simon Bates who used to make the nation sob every morning with 'Our Tune' but I suppose it might be. Whatever, he was one of the middlemen in my supply chain and he worked out that an EORI is some kind of hybrid VAT number. And with that information the seed was freed, and this morning a massive artic lorry hissed to a halt in the farmyard. As I write, I can hear the welcome beeping sound of a reversing telehandler telling me that it's being loaded into the drill.

And that's where I shall be this evening. As the sun

sets on this wonderful spring day, I shall be sitting in the cab with a cold beer trying to spot birds' nests in the blossom-filled hedges as I trundle up and down my fields, planting pasta. Farming has been made even more difficult by Brexit but despite that, I shall be very happy.

May

Farming is the most dangerous job in Britain

As is the way with everything powered by electricity, the automatic door on my new barn broke recently. This meant the enormous pile of wheat in there became an all-you-can-eat 24-hour diner both for the farm's rats and the pheasants I didn't shoot last year because of Covid.

Naturally I needed to get it closed as soon as possible, but to do that I had to access the machinery box, which was about 15ft above my head. And I don't have a ladder. I don't believe in them and I don't trust them. They work in tandem with the cruel mistress that is gravity to ensure that no matter how many precautions you take, you will end up in great pain.

Only recently a chap who works for me was up a wooden stepladder painting a beam when the top rung snapped. This caused him to fall, with one leg on one side of the ladder and one on the other. As both his hands were full of painting stuff, he couldn't brace himself, which meant he broke the next rung down with his testicles, and then the next and then the next until eventually he was at floor level. Incredibly he didn't spill any paint.

I can still hear the shrill, keening noise he made, though, so there's no way I was going to try to unjam my barn door while perched on a long wooden killing machine. Especially when I have a JCB telehandler.

Now I know that rule one in every single health and safety book says you should never, ever use the forks on a telehandler to raise an actual person, but I've never been able to see why. It's mechanical, not electrical, which means it won't suddenly go wrong for no reason. So out I went with Lisa, who's Irish and therefore a natural with construction equipment, and up I went into the roof.

Inevitably the operation was not a success. The barn door remains resolutely open. But on the upside I was not injured. In fact, and I'm touching wood as I write this, I haven't been injured at all in the eighteen months of farming.

My tractor driver had to go to hospital when he got a fertilizer pellet in his ear – don't ask – and another guy nail-gunned his leg to a scaffolding board, but me? I'm still the same shape as I was. There are no new holes. And all my limbs are still attached.

This makes me unusual as farming remains, by far, the most dangerous job in the country. *Farmers Weekly*, my bible on these matters, reported a couple of weeks ago that over the past twelve months more than fifty people died while working in UK agriculture. That's the highest level for twenty-five years and means that around one farmer or farm worker every week says to their family when they leave the breakfast table, 'See you later.' And then doesn't.

It's clear, then, that when you visit the British countryside you are entering the Killing Fields. And remember, this figure doesn't include the huge number who have to walk home carrying the arm they've cut off or those who end up in a straw bale with no legs. Or dignity.

These days, thanks to the advance of Marklism, you aren't allowed to hurt even a soldier's feelings, so they don't have anything like the rate of injury and death that farmers have. No other job does. Not the police. Not reporters or cameramen in war zones. Hell, I bet if you checked there's probably a better survival rate in the Democratic Republic of Congo's cobalt mines.

I can see why, of course. Most farmers use a quad bike of some sort and these things are by far and away the most stupid way of moving around since Sir Clive Sinclair woke up one morning and said, 'I'm going to make an electric slipper.' The fact is this: sooner or later, quad bikes fall over.

The Travellers who live up the road from my farm have many, and on a lovely evening we can hear them jumping over stuff and doing wheelies. Many of my neighbours complain about the noise and the damage being done, but I see no point. Because soon the noise will stop . . .

Other issues. Well, the back of a tractor seems to me to be a particularly dangerous environment. You are usually standing on slippery mud, trying to maintain your balance while attempting to attach lumps of powerful machinery to a surface that's invariably covered in a thick layer of grease and oil. And you're almost always on your own, miles from anywhere. And as often as not it's dark. And raining.

You sometimes watch astronauts attempting to restart broken satellites and you think, 'Ooh, that looks dangerous.' But compared with tractoring, it's like being a schoolteacher in Godalming. Imagine covering yourself in baby oil then playing Twister in a Burmese sawmill. Then you're on the right page.

Another issue in farming is being crushed. I don't even want to think about that, but soon I shall have to because it's that time of year when I have to cultivate my game covers. And one is located on a precipitous slope. Last year I was almost sitting on the side window of my tractor as I went along and I was acutely aware that at any moment it could topple over and that I'd end up underneath it, being pushed into the ground like a tent peg. I'm told this is the most painful way to die.

But I reckon being burnt alive would be worse, and that's another peril I face once a month when the time comes to burn some stuff that won't light because it's a bit damp, so I think I'll use a bit of diesel and it turns out a bit's not enough, so I lob a gallon on to the fire that I think had gone out and then there's a *whoomph*, suggesting that it hadn't.

So far I've only ever lost my eyebrows, but the day can't be far away when I am forced to run around in a burning panic, like one of those flamethrower people in *Saving Private Ryan*.

However, it turns out that fire and falling off ladders and overturning tractors are not the biggest danger out here in the sticks. We read recently that you are just as likely to be killed by a cow as you are by an AstraZeneca-related blood clot, and that heartened us all because, obviously, the only people who are ever killed by cows are men in satin suits in Spain.

Not so, it seems. Cows are incredibly dangerous. Some reports suggest that they are in fact the biggest killer of countryside people in Britain. Not just farmers, but walkers as well.

I was thinking of maybe getting a few Friesians because I like them, but I'm told that a Friesian bull is more dangerous than a saltwater crocodile on a quad bike. Friesian bulls remember if you look at them strangely and can make a mental note to kill you at a later date. They actually bear grudges and their attacks are premeditated. They are black and white terminators.

So bear that in mind the next time you are watching *First Man* or *The Right Stuff*. By all means marvel at the bravery of those test pilots and swoon at the skill, but never forget that compared with a farmer, and a beef farmer in particular, Neil Armstrong and Chuck Yeager were pussies.

Get orf my land!

My rape's gone wrong. It's tufty and sporadic and the insipid hue of powdered custard. Some say this is because it's all been eaten by pigeons. Others blame the ban on insecticides or the phenomenally cold spring we've just had. But these scenarios seem unlikely as, just half a mile away, my neighbouring farmer's crop is vibrant and rich and the lustrous colour of a JCB.

I gaze upon his fields and it's like being at a party with the parents of particularly high-achieving children. You know, your boy is working in JD Sports, selling tracksuits to morons, so you really don't want to hear that 'Rupert's on the International Space Station and Alex has just got cold fusion to work.'

Frankly, I blame myself because in one field the failed areas are in worryingly neat rectangles, suggesting that I pushed the wrong button when I was operating the seed drill. In another, called Bake's Piece – all fields have stupid names – half the acreage is sort of alive and then, south of a dead straight line, it sort of isn't. And pigeons don't eat in straight lines.

In James May Is a Dildo – I may have named that field myself – there's about ten acres of volunteer rape that is growing all by itself. And it's doing a whole lot better than the stuff that I've nurtured carefully these past few months.

The only good news is that rape prices are way up this year. Because of the Europe-wide ban on neonicotinoid pesticides – sprays that kill bees – many farmers haven't bothered planting it. So demand will exceed supply.

I may only grow an ounce but it could well fetch more than a handful of diamonds.

This is a worry because if locally grown oil-seed rape is expensive, vegetable oil will cost more. Which means consumers will be more inclined to buy palm oil instead. That would be bad.

In the not-too-distant past Sir Attenborough made a show about some orangutans that live in Sumatra. There was one heart-stopping scene where they had to teach their youngsters how to swing over a river that was full of crocodiles, and I couldn't help thinking: 'Why are they doing that? Why don't they stay on their side of the river until the toddlers are less likely to fall into the water?'

Then we got a drone shot that brought the problem into sharp focus. The jungle extended only a few hundred yards from the riverbanks, and after that it was gone, replaced by an endless and completely orangutan-unfriendly forest of palm trees. It was a man-made farm, built so that thick and thin 'eco-conscious' people in the West can have fresh, sweet-smelling faces.

These people do not like the vegetable oil that comes from my rape. They even say it makes them sneeze. They'd rather kill an orangutan for a bit of imported palm oil. And there's a very good reason. They don't know anything about where anything comes from. They sit, over an avocado breakfast, telling their friends they want locally produced, carbon-neutral food, but it's simply not true. They just want it to be cheap.

This is why I'm such a fan of the Duke of Edinburgh's Award scheme. Because it takes kids who would otherwise spend their weekends in suburban bus shelters into the countryside where, hopefully, they learn about where the bread in their Subway comes from.

Earlier this month I found a gang of them sitting in my garden. Some were attempting to light a stove, some were moaning about the mobile phone signal, and one was heading for my hedge with some bog roll. Naturally I said, 'Can I help?', which is farmer-speak for 'get off my land'.

It turned out they were on a DofE course and had been told by their teacher, who was doubtless a jumped-up little socialist who believes that property is theft, that they should have their lunch in the garden of the biggest house they could find. It also turned out that the girl with the bog roll needed 'to go toilet'.

'Well, you can't defecate in my hedge,' I said. 'You're not an animal. So let's go to the farm where you can use the loo.'

At this point some kind of supervisor arrived to say that while it was OK for him to spend his free time, in the woods, with a load of prepubescent girls, it was emphatically not OK for me to take one of them to the lavatory.

She would have to be accompanied by a 'buddy', who may or may not have also been a girl. But before I'd had a chance to think of what pronoun to use, there was the Covid problem of how best I could drive them to the farm when we had to be 6ft apart and in hazmat suits. And while that was being sorted out, the poor girl who needed to go toilet was turning maroon.

I then took a call from my tractor driver to say that the supervisor's Peugeot was blocking the track so I asked him to move it. Whereupon he summoned his 'I know my rights' adenoids to say that it was a public footpath, and I had to retort in my special menacing tone, explaining that you can't drive on a footpath, or picnic on it, and that you sure as hell can't defecate on it.

With toilet girl now looking like a U-boat's boiler, I

received word that another group of cisgenderists had descended on a field called Quarry – because it doesn't have a quarry in it. The field that does have a quarry in it is called Banks.

So I threw Covid caution to the wind, dumped the human pressure cooker outside the farm bog and raced down the track, past all the other abandoned Peugeots to see what was what.

The scene was alarming. Because in a strip I've just planted with an extremely expensive turtle dove mix, it looked like some kind of juvenile scat movie was in full swing. They were kids, I know that, and they were out of their comfort zone. But they were also pushing used lavatory paper into the watering pipes for my new trees. So I was calm but firm when I asked where their supervisor was.

It turned out he was in the village, and that's the problem so far as I can tell. Because while it's great that these kids are out in the fresh air, learning how to get around using nothing but abandoned teacher Peugeots as marker points, it doesn't really help them understand the land if there's no one around to tell them stuff.

I desperately wanted to explain my thinkings on palm oil and the need to buy vegetable oil instead, but I don't think these Kardashian and Snapchat enthusiasts would have quite got it. Besides, at some point I'd have had to use the word 'rape', which would have meant me writing this in a cell.

As a result they're now back at school and what have they learnt? That the mobile phone coverage in the countryside is dismal, that it's cold and that everyone who lives there is fat and angry.

Postscript

Ending a book like that, suddenly and with no conclusion, is weird. It'd be like finishing the *Day of the Jackal* as the assassin took his gun out of his crutch. But that's farming. Out here in the fields, Michael Lonsdale does not burst through the door with a machine gun. The story just goes on. And on.

We like to think that the harvest is some kind of end but it isn't. Because long before the last grain lorry has left the yard, the farmer is back in the fields, getting them ready for next year.

I did think when I had finished a year on the farm that I might go back to London and spend the rest of time eating stuff that other people had made. But it was a lovely September day and the leaves in the trees were beginning to go orange and brown and red and I looked back on what I'd learned over the previous twelve months and knew that I was going to stick around. Even though I was only being paid forty pence a day.

I have never taken anything particularly seriously; life doesn't seem much worth living unless it's amusing, but farming has made me realize that you can still have fun whilst doing something sensible and worthwhile.

So, there we are. I love writing about farming, but I love doing it even more.

Jeremy 'farmer' Clarkson

He just wanted a decent book to read ...

Not too much to ask, is it? It was in 1935 when Allen Lane, Managing Director of Bodley Head Publishers, stood on a platform at Exeter railway station looking for something good to read on his journey back to London. His choice was limited to popular magazines and poor-quality paperbacks – the same choice faced every day by the vast majority of readers, few of whom could afford hardbacks. Lane's disappointment and subsequent anger at the range of books generally available led him to found a company – and change the world.

'We believed in the existence in this country of a vast reading public for intelligent books at a low price, and staked everything on it'
Sir Allen Lane, 1902–1970, founder of Penguin Books

The quality paperback had arrived – and not just in bookshops. Lane was adamant that his Penguins should appear in chain stores and tobacconists, and should cost no more than a packet of cigarettes.

Reading habits (and cigarette prices) have changed since 1935, but Penguin still believes in publishing the best books for everybody to enjoy. We still believe that good design costs no more than bad design, and we still believe that quality books published passionately and responsibly make the world a better place.

So wherever you see the little bird – whether it's on a piece of prize-winning literary fiction or a celebrity autobiography, political tour de force or historical masterpiece, a serial-killer thriller, reference book, world classic or a piece of pure escapism – you can bet that it represents the very best that the genre has to offer.

Whatever you like to read – trust Penguin.